Electronics III
A textbook covering the Level III syllabus
of the Technician Education Council

D C Green
M Tech, CEng, MIERE
Senior Lecturer in Telecommunication Engineering
Willesden College of Technology

Second edition

Pitman

PITMAN BOOKS LIMITED
128 Long Acre, London WC2E 9AN

Associated Companies
Pitman Publishing Pty Ltd, Melbourne
Pitman Publishing New Zealand Ltd, Wellington

© D. C. Green 1982

First published in Great Britain 1978
Reprinted 1980, 1981, 1982
Second edition 1982
Reprinted 1983

Text set in 10/12 Linotron Times,
Printed and bound in Great Britain
at The Pitman Press, Bath

ISBN 0 273 01831 0

Contents

Preface to the second edition

The second edition of this book takes account of the changes that have been made to the TEC unit Electronics III. This has meant the introduction of some material on bipolar transistors into Chapter 1, and on Schmitt triggers and sawtooth generators into Chapter 8. A number of other small changes have also been made at various points in the text. Appendix A, which gives outline details of some linear integrated circuits, is also new.

Chapter 1 covers the requirements of the unit on the topic of field effect transistors but much of its content also appears in another volume, i.e. Electronics II. This has been done because it was felt that some colleges would like to introduce their students to the field effect transistor at the second level.

D.C.G.

Preface to the first edition

This book provides a comprehensive coverage of the electronic circuits and techniques employed in modern analogue electronic and telecommunication equipment. A good knowledge of electronic techniques is an essential part of the education of a telecommunication technician whether his interests lie mainly in the field of line or radio engineering or in telephone switching.

The Technician Education Council (TEC) scheme for the education of electronic/telecommunication technicians introduces the basic principles of electronics in the second-level unit Electronics II and further understanding of electronics is provided in the third-level unit Electronics III. This book has been written to provide a complete coverage of the contents of the Electronics III unit.

Chapters 1 and 2 introduce the basic principles of field-effect transistors and monolithic integrated circuits. In the following chapters the applications of these devices are combined with the bipolar transistor versions. The use of integrated circuits for various functions has been covered in a general manner and does not refer to particular devices in current usage. This approach to integrated circuits has been adopted in an attempt to treat the device as one available means of performing a given circuit function and which requires certain components, such as inductors and capacitors, provided externally. Chapter 3 to 6 inclusive discuss the ways in which the bipolar and field-effect transistors and the integrated circuit can be used for the amplification of signals. Chapter 3 covers small-signal audio-frequency amplifiers while Chapter 4 deals with the application of negative feedback to a.f. amplifiers as well as introducing the operational amplifier. Chapter 5 deals with the use of a.f. power amplifiers and considers both single-ended Class A and push-pull Class B circuits; integrated power amplifiers are also mentioned. The use of tuned amplifiers at radio frequencies to provide both gain and selectivity is the subject of Chapter 6.

The next two chapters, that is 7 and 8, deal with the methods commonly employed to generate waveforms. Sinusoidal oscillators using both L-C and R-C networks and piezoelectric crystals are described in Chapter 7. Chapter 8 is concerned with the various kinds of multivibrator available and shows the ways in which each type can be made. Chapters 9, 10 and 11 cover, respectively, the sources of noise in electronic and communication systems, the differentiation and integration of waveforms, and stabilized power supplies for electronic equipment.

The prior knowledge needed to use the book is only that which should already be possesed by a technician who has completed a course covering the level II subject Electronics II. In particular, a knowledge of the principles of operation of the bipolar transistor has been assumed.

The book provides a text that should prove eminently suitable for all students of electronic engineering requiring a mainly non-mathematical approach and in particular for those students undertaking a course of study designed for the TEC unit Electronics III.

Acknowledgment is due to the Technician Education Council for permission to use the content of the TEC unit in the appendix to this book. The Council reserve the right to amend the content of its unit any time.

Many worked examples are provided in the text to illustrate the principles that have been discussed and each chapter concludes with a number of short exercises and longer exercises. Many of the exercises have been taken from past City and Guilds examination papers and grateful acknowledgement of permission to do so is made to the Institute. Answers to the numerical exercises will be found at the rear of the book; these answers are the sole responsibility of the author and are not necessarily endorsed by the Institute.

D.C.G.

1 Transistors

Introduction

Transistors are semiconductor devices used for a wide variety of electronic purposes but mainly for either amplification or switching. Two kinds of transistor are available, in both discrete and integrated circuit versions: the BIPOLAR TRANSISTOR and the FIELD EFFECT TRANSISTOR.

The principles of operation of the bipolar transistor have already been studied [EII] and in this chapter the discussion will be limited to the input and output characteristics of the device and to its h parameters.

The field effect transistor operates in a fundamentally different way from the bipolar transistor and its operation will be considered in this chapter. Two types of f.e.t. are available: the junction f.e.t. (j.f.e.t.) and the metal oxide semiconductor f.e.t. (m.o.s.f.e.t.). Further, the m.o.s.f.e.t. can be sub-divided into two classes: the depletion type and the enhancement type.

The Bipolar Transistor

The operation of the bipolar transistor when it is used as an amplifier can be understood with the aid of its *input* and *output characteristics* and/or its *h parameters*. The use of characteristics for this purpose is demonstrated in Chapters 3 and 5 but gain calculations using the h parameters are outside the scope of this book [see EIV]. The input characteristics of a bipolar transistor show how the base current varies with change in the base/emitter voltage, with the collector/emitter voltage held constant. Fig. 1.1 shows a typical set of input characteristics for a BC 107 silicon planar transistor when the collector/emitter voltage is 6 V. Since this is a silicon device, no base current flows until the base/emitter voltage V_{BE} is greater than about 0.5 V and the base current only starts to

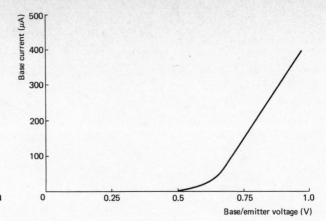

Fig. 1.1 Input characteristics of a BC 107 bipolar transistor

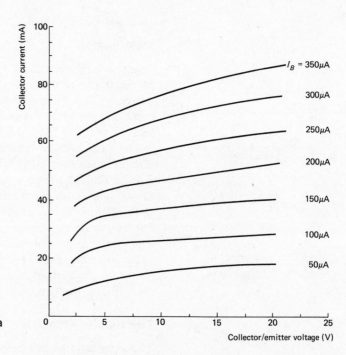

Fig. 1.2 Output characteristics of a BC 107 bipolar transistor

increase more or less linearly when V_{BE} exceeds about 0.6 V.

The output characteristics of a bipolar transistor indicate the way in which the collector current varies with change in the collector/emitter voltage, with the base current held constant. A family of curves is generally drawn to show the effect of different values of base current. Fig. 1.2 shows a typical set of output characteristics for a BC 107 transistor. Although collector currents up to nearly the maximum current rating of 100 mA are shown, the transistor is most often operated with collector currents of less than 10 mA.

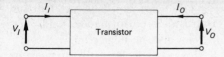

I_I I_O

V_I Transistor V_O

Fig. 1.3 The bipolar transistor as a 4-terminal network

The h parameters of a transistor are defined by considering the transistor to be a four-terminal network (Fig. 1.3). If the input current I_I and the output voltage V_O are taken to be the independent variables, then the input voltage V_I and the output current I_O can be written down as

$$V_I = I_I h_I + V_O h_R \qquad (1.1)$$

$$I_O = I_I h_F + V_O h_O \qquad (1.2)$$

The meaning and the dimension of each of the four h parameters can be determined by setting first V_O, and then I_I, to zero.

With $V_O = 0$, equation (1.1) becomes $V_I = I_I h_I$. Therefore,

$$h_I = V_I/I_I \, \Omega \qquad (1.3)$$

h_I is the D.C. INPUT RESISTANCE in ohms with the output terminals short-circuited.

Also, from equation (1.2), $I_O = I_I h_F$ or

$$h_F = I_O/I_I \text{ (no units)} \qquad (1.4)$$

h_F is the FORWARD D.C. CURRENT GAIN with the output terminals short-circuited.

With $I_I = 0$, i.e. with the input terminals open-circuited, equations (1.1) and (1.2) become

$$V_I = V_O h_R \quad \text{and} \quad I_O = V_O h_O \quad \text{respectively}$$

Therefore,

$$h_R = V_I/V_O \text{ (no units)} \qquad (1.5)$$

h_R is the REVERSE D.C. VOLTAGE RATIO with the input terminals open-circuited.

Finally

$$h_O = I_O/V_O \text{ S} \qquad (1.6)$$

h_O is the D.C. OUTPUT CONDUCTANCE in siemens with the input terminals open-circuited.

These results are true for *any* four-terminal network. When they are applied to a bipolar transistor, a second suffix is added to indicate the configuration in which the transistor is connected. Thus, for the common-emitter connection the four h parameters are

$$h_{IE}, \; h_{FE}, \; h_{RE}, \; h_{OE}$$

The use of the h parameters is nearly always associated with the calculation of circuit performance under *small-signal* conditions. Then each of the four currents and voltages shown in Fig. 1.3 are *alternating quantities* and the symbols used for the h parameters are

$$h_{ie}, \; h_{fe}, \; h_{re}, \; h_{oe} \qquad \text{i.e. lower-case suffices are used.}$$

For modern transistors the parameter h_{re} is always small and has little effect on the transistor's performance, and it is usually neglected. Often h_{oe} can also be neglected.

The values of h_{IE} and h_{ie} for a particular transistor can be determined from its input characteristic using equation (1.3). The output characteristic of a transistor can be used to determine the values of h_{OE}, h_{oe}, h_{FE} and h_{fe}. In all cases, when the small-signal parameters are required, *changes* in voltage and/or current are involved. This means that h_{ie} is the *slope* of the input characteristic and h_{oe} is the *slope* of the output characteristic at the point of measurement.

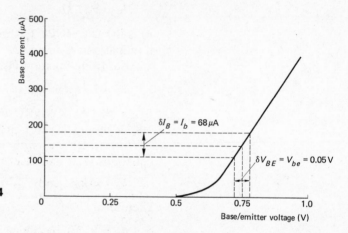

Fig. 1.4

Consider Fig. 1.4 which shows the input characteristic of the BC 107 again and suppose that both h_{IE} and h_{ie} are to be determined at the point where the base/emitter voltage V_{BE} is 0.75 V. Then, h_{IE} is the ratio V_{BE}/I_B for *static* values. Hence from the graph (dotted lines)

$$h_{IE} = 0.75/145 \times 10^{-6} = 5172 \ \Omega$$

h_{ie} is the ratio $\delta V_{BE} \,|\, \delta I_B = V_{be}/I_b = 0.05/68 \times 10^{-6} = 735 \ \Omega$. It is evident that considerable difference exists between the values of h_{IE} and h_{ie}. If the point of measurement is reduced below about $V_{BE} = 0.6$ V, the slope of the input characteristic is smaller and the a.c. input resistance h_{ie} will be larger so that the difference between h_{IE} and h_{ie} will be much less. For base currents of the order of a few tens of microamps, h_{ie} is typically about 3 kΩ.

The output characteristic of the BC 107 is redrawn in Fig. 1.5; h_{OE} is the ratio I_C/V_{CE} and h_{oe} is the ratio $\delta I_C/\delta V_{CE} = I_c/V_{ce}$. The values of these parameters are to be measured at the point $V_{CE} = 12.5$ V.

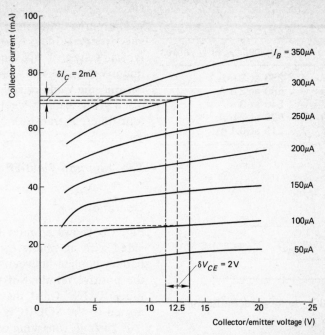

Fig. 1.5

From the dotted lines, for the curve $I_B = 100 \ \mu$A, $h_{OE} = 27 \times 10^{-3}/12.5 = 2.16$ mS and, for the curve $I_B = 300 \ \mu$A, $h_{OE} = 69 \times 10^{-3}/12.5 = 5.52$ mS. It is obvious that h_{OE} must increase with increase in the collector current.

From the chain-dotted lines, when $I_B = 300 \ \mu$A, then $h_{oe} = 2 \times 10^{-3}/2 = 1$ mS. When $I_B = 100 \ \mu$A, the change in I_B produced by a 2 V change in V_{CE} is too small to measure on the graph; this does mean, of course, that h_{oe} decreases with decrease in the steady collector current, and is typically about $25 \ \mu$S for collector currents of about 2 mA.

The parameters h_{FE} and h_{fe} can also be measured from the output characteristics of Fig. 1.5. If the collector/emitter voltage is kept constant at 12.5 V, the collector current will increase from 27 mA to 69 mA as the base current is increased from $100 \ \mu$A to $300 \ \mu$A. Hence

$$h_{fe} = (69-27) \times 10^{-3}/(300-100) \times 10^{-6} = 210$$

Also, at $I_B = 100 \ \mu$A,

$$h_{FE} = 27 \times 10^{-3}/100 \times 10^{-6} = 270$$

and at $I_B = 300 \ \mu$A

$$h_{FE} = 69 \times 10^{-3}/300 \times 10^{-6} = 230$$

Although in this case $h_{fe} < h_{FE}$ the difference between the d.c. and small-signal values of short-circuit current gain depends upon the operating conditions chosen. In most cases little error is introduced by assuming that h_{fe} is equal to h_{FE}.

Table 1.1

Type	Typical h_{FE} at I_C values
BC 107	290 at 2 mA
BC 159	650 at 2 mA
BC 177	240 at 3 mA
BC 109	520 at 2 mA
BC 154	215 at 0.1 mA

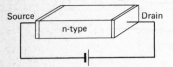

Fig. 1.6 n-type semiconductor

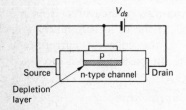

Fig. 1.7 The basic junction f.e.t.

The parameters of individual transistors of the same type vary from one device to another. Manufacturer's data quote typical values of h_{FE} and very often minimum and/or maximum values as well. Since h_{FE} varies with the d.c. collector current, the value of collector current for the maximum h_{FE} is often given also. Table 1.1 gives some typical h_{FE} figures for four different transistors.

The Junction Field-Effect Transistor

Operation

Fig. 1.6 shows a wafer of lightly-doped n-type silicon, provided with an ohmic contact at each of its two ends, and a battery applied between these contacts. The contact to which the positive terminal of the battery is connected is known as the DRAIN, whilst the negative side of the battery is connected to the SOURCE contact.

A current, consisting of majority charge carriers, will flow in the silicon wafer from drain to source, the magnitude of which is inversely proportional to the resistance of the wafer. This current is known as the DRAIN CURRENT. The resistance of the wafer in turn depends upon the resistivity of the n-type silicon wafer and the length and cross-sectional area of the conducting path, or *channel*, i.e. $R = \rho l / a$. For given values of resistivity and length, the channel resistance will depend upon the cross-sectional area of the channel. If, therefore, the cross-sectional area can be varied by some means, the channel resistance and hence the drain current can also be varied.

The properties of a p-n junction are such that the region either side of the junction, known as the "depletion layer," is a region of high resistivity whose width is a function of the reverse-biased voltage applied to the junction. The depletion layer can be used to effect the required control of the channel resistance. A p-n junction is therefore required in the silicon wafer and to obtain one it is necessary to diffuse a p-type region into the wafer, as shown by Fig. 1.7. The p-type region is doped more heavily than the n-type channel to ensure that the depletion layer will lie mainly within the channel. An ohmic contact is provided to the p-type region and it is known as the gate terminal.

If the gate is connected directly to the source, the p-n junction will be reverse biased and the depletion layer will be extended further into the channel. The p-n junction is reverse biased because the p-type gate region is at zero potential, while the n-type channel region is at some positive potential. A potential gradient will exist along the length of the channel,

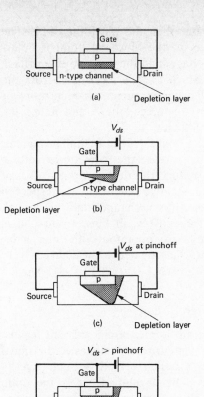

(a)

V_{ds}

(b)

V_{ds} at pinchoff

(c)

V_{ds} > pinchoff

(d)

Fig. 1.8 Showing the effect of increasing the drain-source voltage

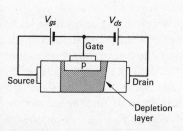

Fig. 1.9 Showing the effect of increasing the gate-source voltage

varying from a positive value equal to the battery voltage at the drain end to zero voltage at the source end. Since the cross-sectional area of the channel between the gate region is smaller than at either end of the channel (because of the depletion layer), the resistance of this area is relatively large, and most of the voltage drop appears across this part of the channel. The drain end of the channel lying in between the gate region is at a higher potential than the source end of the channel; hence the reverse-bias applied to the p-n junction is greater on this side. The effect on the depletion layer is shown in Fig. 1.8.

When the drain-source voltage is zero (Fig. 1.8*a*), the depletion layer either side of the p-n junction is narrow and has little effect on the channel resistance. Increasing the drain-source voltage above zero will widen the depletion layer and cause it to extend into the channel. This is shown in Fig. 1.8*b* which makes it clear that the layer widens more rapidly at the drain end of the channel than at the source end.

Thus, increasing the drain-source voltage increases the channel resistance and this results in the increase in the drain current being less than proportional to the voltage, i.e. doubling the drain-source voltage does not give a two-fold increase in drain current because the channel resistance has increased also. Further increase in the drain-source voltage makes the depletion layer extend further into the channel and eventually the point is reached where the depletion layer extends right across the channel (Fig. 1.8*c*). The drain-source voltage which produces this effect is known as the PINCH-OFF VOLTAGE. Once pinch-off has developed, further increase in drain-source voltage widens the pinched-off region (Fig. 1.8*d*). The drain current ceases to increase in proportion to any increase in the drain-source voltage, but is now more or less constant with change in drain-source voltage. The drain current continues to flow because a relatively large electric field is set up across the depleted region of the channel, and this field aids the passage of electrons through the region.

The reverse-bias voltage applied to the gate-channel p-n junction can also be increased by the application of a negative potential, relative to source, to the gate terminal. If the gate-source voltage is made negative, the reverse bias on the gate-channel junction is increased. This increase in bias voltage widens the depletion layer over the width of the gate region and thereby increases the channel resistance. The drain current therefore falls as the gate-source voltage is made more negative until it is approximately equal to the pinch-off voltage and the channel is pinched-off (Fig. 1.9). When this occurs the drain current is zero. Generally, the junction f.e.t. is operated with voltages applied to both the drain and the gate terminals,

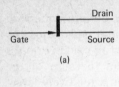

(a)

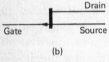

(b)

Fig. 1.10 Symbols for (a) an n-channel j.f.e.t. and (b) a p-channel j.f.e.t.

with the drain voltage greater than the pinch-off value. The resistance of the channel up to the pinch-off point is determined by the gate-source voltage, and the drain-source voltage produces an electric field which sweeps electrons across the extended depletion layer. The drain current is then more or less independent of the drain-source voltage and under the control of the gate-source voltage.

A p-channel junction f.e.t. operates in a similar manner except that it is necessary to increase the gate-source voltage in the positive direction to reduce the drain current. Also, of course, the drain is held at a negative potential with respect to the source. The symbols used for n-channel, and p-channel, junction f.e.t.s are given, respectively, in Figs. 1.10a and b. Both types of junction f.e.t. are operated with their gate-channel p-n junction reverse biased; hence they have a very high input impedance.

Application in an Amplifier Circuit

If a junction f.e.t. is to operate in an amplifier circuit it must be possible to control the drain current by means of the signal voltage. If the drain current is then passed through a resistance, an output voltage will be developed across the drain resistance that is an amplified version of the input signal voltage. The necessary control of the drain current can be obtained by connecting the signal voltage in the gate-source circuit of the f.e.t. (Fig. 1.11).

The signal source, of e.m.f. E_s and impedance R_s, is connected in the gate-source circuit of the f.e.t. in series with a bias battery of e.m.f. V_{gs}. The total reverse bias voltage applied to the gate-channel junction is the sum of the signal voltage E_s, and the bias voltage V_{gs}. During the positive half-cycles of the signal waveform, the reverse junction bias is reduced, the depletion layer becomes narrower, and so the drain current increases. Conversely, negative half-cycles of the signal waveform augment the bias voltage and cause the depletion layer to extend further into the channel; the drain current is therefore reduced. In this way, the drain current is caused to vary with the same waveform as the input signal.

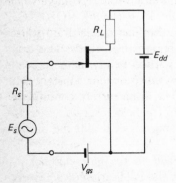

Fig. 1.11 The basic j.f.e.t. amplifier

The output voltage is developed across the drain load resistor R_L, and can be taken off from between the drain and earth. A voltage gain is achieved because the alternating component of the voltage across R_L is larger than the signal voltage E_s. An increase in the signal voltage in the positive direction produces an increase in the drain current and hence an increase in the voltage developed across R_L. The drain-source voltage V_{ds} is the difference between the drain supply voltage E_{dd} and the voltage across R_L; thus an increase in drain current makes the drain-source voltage fall. This means that a junction f.e.t. amplifier operated in the common-source configuration has its input and output signal waveforms in antiphase with one another. It is necessary to ensure that the signal voltage is not large enough to take the gate-source voltage positive by more than about 0.5 V, otherwise the high input-impedance feature of the junction f.e.t. will be lost.

The CONSTRUCTION of a junction f.e.t. is shown in Fig. 1.12g.

The various steps involved in the manufacture of an n-channel junction f.e.t. are shown by Figs. 1.12a through to g. A heavily-doped p-type silicon substrate marked as p^+ in Fig. 1.12a has a layer of silicon dioxide grown onto its surface (Fig. 1.12b). Next (Fig. 1.12c) a part of the silicon dioxide layer is etched away to create an exposed area of the p-type silicon substrate into which n-type impurities can be diffused. An n-type region is thus produced in the p-type substrate and then another layer of silicon dioxide is grown onto the surface (Fig. 1.12d). The next steps, shown by Fig. 1.12e, are first to etch another gap in the silicon dioxide layer and then to diffuse a p^+ region into the exposed area of the n-type region of the substrate. A third layer of silicon dioxide is then grown over the surface of the device (Fig. 1.12f). Gaps are now etched into the layer into which aluminium contacts to the two ends of the n-type region and the upper p-type region can be deposited (Fig. 1.12g). The terminals connected to the two ends of the n-type region are the source and the drain contacts, while the third terminal acts as the gate.

The important parameters of a junction f.e.t. are its mutual conductance g_m, its input resistance R_{IN}, and its drain-source resistance r_{ds}. The mutual conductance is defined as the ratio of a change in the drain current to the change in the gate-source voltage producing it, with the drain-source voltage maintained constant, i.e.

$$g_m = \frac{\delta I_d}{\delta V_{gs}} \qquad V_{ds} \text{ constant} \qquad (1.7)$$

The drain-source resistance r_{ds} is the ratio of a change in the drain-source voltage to the corresponding change in drain current, with the gate-source voltage held constant, i.e.

$$r_{ds} = \frac{\delta V_{ds}}{\delta I_d} \qquad V_{gs} \text{ constant} \qquad (1.8)$$

Typically, g_m has a value lying in the range of 1 to 7 mS, while r_{ds} may be 40 kΩ to 1 MΩ. The input impedance of a junction f.e.t. is the high value presented by the reverse-biased gate-channel p-n junction. Typically, an input impedance in excess of 10^8 Ω may be anticipated.

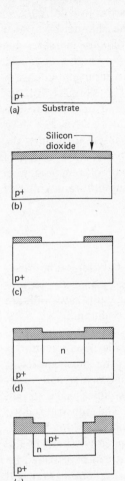

Fig. 1.12 The stages in the manufacture of an n-channel j.f.e.t.

The M.O.S.F.E.T.

The metal-oxide-semiconductor field-effect transistor, generally known as the m.o.s.f.e.t., differs from the junction f.e.t. in that its gate terminal is insulated from the channel by a layer of silicon dioxide. The layer of silicon dioxide increases the input impedance of the f.e.t. to an extremely high value, such as $10^{10}\ \Omega$ or even more. The high value of input impedance is maintained for all values and polarities of gate-source voltage, since the input impedance does not depend upon a reverse-biased p-n junction.

The m.o.s.f.e.t. is available in two different forms: the depletion type and the enhancement type. Both types of m.o.s.f.e.t. can be obtained in both n-channel and p-channel versions, so that there are altogether four different kinds of m.o.s.f.e.t.

Depletion-type M.O.S.F.E.T.

The constructional details of an n-channel depletion mode m.o.s.f.e.t. are shown in Fig. 1.13. Two heavily doped n^+ regions are diffused into a lightly doped p-type substrate and are joined by a relatively lightly-doped n-type channel.

The gate terminal is an aluminium plate that is insulated from the channel by a layer of silicon dioxide. A connection is also made via another aluminium plate to the substrate itself. In most m.o.s.f.e.t.s the substrate terminal is internally connected to the source terminal but sometimes an external substrate connection is made available. The substrate must always be held at a negative potential relative to the drain to ensure that the channel-substrate p-n junction is held in the reverse-biased condition. This requirement can be satisfied by connecting the substrate to the source. A depletion layer will extend some way into the channel, to a degree that depends upon the magnitude of the drain-source voltage. Because of the voltage dropped across the channel resistance by the drain current, the depletion layer extends further across the part of the channel region nearest to the drain than across the part nearest the source. The resistance of the channel depends upon the depth to which the depletion layer penetrates into the channel. With zero voltage applied to the gate terminal the drain current will, at first, increase with increase in the drain-source voltage, but once the depletion layer has extended right across the drain end of the channel the drain current becomes, more or less, constant with further increase in the drain-source voltage.

The channel resistance, and hence the drain current, of a depletion-type m.o.s.f.e.t. can also be controlled by the voltage

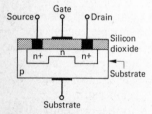

Fig. 1.13 Construction of an n-channel depletion-type m.o.s.f.e.t.

applied to the gate. A positive voltage applied to the gate will attract electrons into the channel from the heavily-doped n$^+$ regions at either end. The number of free electrons available for conduction in the channel is increased and so the channel resistance is reduced. The reduction in channel resistance will, of course, allow a larger drain current to flow when a given voltage is maintained between the drain and source terminals. An increase in the positive gate voltage will increase the drain current which flows when the drain-source voltage is large enough to extend the depletion layer across the drain end of the channel. Conversely, if the gate is held at a negative potential relative to the source electrons are repelled out of the channel into the n$^+$ regions. This reduces the number of free electrons which are available for conduction in the channel region and so the channel resistance is increased. The drain current that flows when the depletion layer has closed the channel depends upon the channel resistance.

The drain current of a depletion-type m.o.s.f.e.t. can therefore be controlled by the voltage applied between its gate and source terminals.

Enhancement-type M.O.S.F.E.T.

Fig. 1.14 shows the construction of an enhancement type m.o.s.f.e.t. The gate terminal is insulated from the channel by a layer of silicon dioxide, and the substrate and source terminals are generally connected together to maintain the channel-substrate p-n junction in the reverse-biased condition. It can be seen that a channel does not exist between the n$^+$ source and drain regions; hence the drain current that flows when the gate-source voltage is zero is very small. If, however, a voltage is applied between the gate and source terminals, which makes the gate positive with respect to the source, a *virtual channel* will be formed. The positive gate voltage attracts electrons into the region beneath the gate to produce an n-type channel (as shown in the figure) in which a drain current is able to flow. The positive voltage that must be applied to the gate to produce the virtual channel is called the *threshold voltage* and is typically about 2 V. Once the virtual channel has been formed, the drain current which flows depends upon the magnitude of both the gate-source and drain-source voltages. An increase in the gate-source voltage above the threshold value will attract more electrons into the channel region and will therefore reduce the resistance of the channel. The drain current produces a voltage drop along the channel and as with the other types of f.e.t., pinch-off will occur for a particular value of drain voltage. For a particular value of gate-source voltage the drain current will increase with increase in drain-

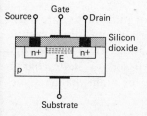

Fig. 1.14 Construction of an n-channel enhancement-type m.o.s.f.e.t. IE = Induced electrons forming a virtual channel when gate voltage is positive

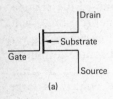

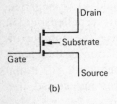

Fig. 1.15 Symbols for (a) an n-channel depletion-type m.o.s.f.e.t. and (b) an n-channel enhancement-type m.o.s.f.e.t.

source voltage up to onset of pinch-off and thereafter will remain more or less constant.

The drain current of a m.o.s.f.e.t. can hence be controlled by the voltage applied between its gate and source terminals and, if the drain current is passed through a suitable resistance, a voltage gain can be provided. The basic arrangement of a m.o.s.f.e.t. amplifier is similar to the junction f.e.t. circuit given in Fig. 1.11 and it operates in a similar manner.

The important parameters of a m.o.s.f.e.t. are the same as those of a junction f.e.t.: namely, its mutual conductance g_m, its drain-source resistance r_{ds}, and its input resistance R_{IN}. Typically, g_m is in the range 1–10 mS, r_{ds} is some 5–50 kΩ, and R_{IN} is 10^{10} Ω or more. It should be noted that whereas the values of mutual conductance are approximately the same as those of a junction f.e.t., the drain-source resistance values are lower but the input resistance is higher.

Figs. 1.15a and b show the symbols for n-channel depletion type and enhancement-type m.o.s.f.e.t.s. The symbols for the p-channel versions differ only in that the direction of the arrow-head is reversed.

Static Characteristics

The static characteristics of a f.e.t. are plots of drain current against voltage and are used to determine the drain current which flows when a particular combination of gate-source and drain-source voltage are applied. Two sets of static characteristics are generally drawn: these are the drain characteristics and the mutual characteristics.

Drain Characteristics

The drain characteristics of a f.e.t. are plots of drain current against drain-source voltage for constant values of gate-source voltage. The characteristics can be determined with the aid of a circuit such as that shown in Fig. 1.16 for the measurement of the characteristics of an n-channel junction f.e.t.

The data required to plot the drain characteristics consists of the values of the drain current which flows as the drain-source

Fig. 1.16 Circuit for the determination of the static characteristics of an n-channel j.f.e.t.

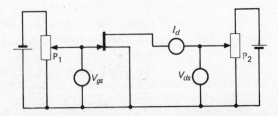

voltage is increased in a number of discrete steps starting from zero, the gate-source voltage being held constant at a convenient value. The method generally used to obtain the data is as follows: the gate-source voltage is set to a convenient value by means of the potential divider P_1 and then the drain-source voltage is increased, starting from zero, in a number of discrete steps. At each step the drain current flowing is noted. The gate-source voltage is then set to another convenient value and the procedure is repeated. In this way sufficient data can be obtained to plot a family of curves of drain current to a base of the drain-source voltage. This family of curves is known as the drain characteristics of the f.e.t. The drain characteristics of the other types of f.e.t. are obtained in a similar manner. Fig. 1.17 shows typical drain characteristics for the six types of f.e.t.

It should be noted that each curve has a region of small values of V_{ds} in which I_d is proportional to V_{ds}. In these

Fig. 1.17 The drain characteristic of
(a) an n-channel j.f.e.t.,
(b) a p-channel j.f.e.t.,
(c) an n-channel depletion type m.o.s.f.e.t.,
(d) a p-channel depletion type m.o.s.f.e.t.,
(e an n-channel enhancement-type m.o.s.f.e.t.,
(f) a p-channel enhancement-type m.o.s.f.e.t.

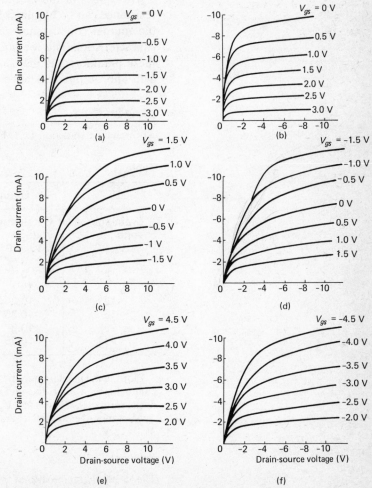

regions the devices can be operated as a voltage-dependent resistance, i.e. as a resistance whose value V_{ds}/I_d depends upon the value of V_{ds}. For all devices the drain current which flows when the gate-source voltage is zero is labelled as I_{dss}.

Mutual Characteristics

The mutual or transfer characteristics of a f.e.t. are plots of drain current against gate-source voltage for various constant values of drain-source voltage. The mutual characteristics of an n-channel junction f.e.t. can be determined using the arrangement given in Fig. 1.16 and the following procedure. The drain-source voltage is maintained at a constant value as the gate-source voltage is increased in a number of discrete steps. At each step the value of the drain current flowing is noted. The procedure should then be repeated for a number of other drain-source voltages.

The values of the mutual conductance and the drain-source resistance can be obtained from the drain characteristics, while the mutual conductance can be determined from the mutual characteristics. The method employed to obtain the values of these parameters is the same as that to determine the current gain and output resistance of a bipolar transistor.

EXAMPLE 1.1

An n-channel junction f.e.t. has the data given in Table 1.1.

Table 1.1

Drain-source voltage V_{ds}(V)	Drain current (mA)			
	Gate-source voltage $V_{gs} = 0$ V	$= -1$ V	$= -2$ V	$= -3$ V
0	0	0	0	0
4	0	5.0	2.4	0.30
8	10.1	5.9	2.7	0.35
12	10.2	6.2	2.9	0.40
16	10.25	6.3	3.0	0.45
20	10.3	6.35	3.05	0.50
24	10.35	6.4	3.1	0.55

Plot the drain characteristics and use them to determine the mutual conductance g_m of the device at $V_{ds} = 12$ V. Calculate also the drain-source resistance for $V_{gs} = -2$ V.

Also plot the mutual characteristics and from them obtain g_m at $V_{ds} = 12$ V.

Solution

The drain characteristics of the f.e.t. are shown in Fig. 1.18.

dc LoadLine Vds = 24V

ac LoadLine

Operating point is at the junction of both load lines

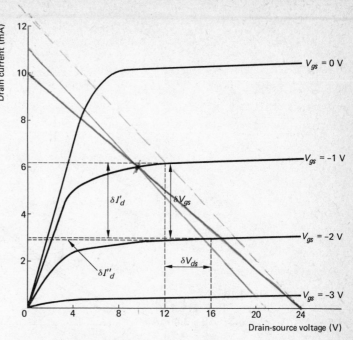

Fig. 1.18
$\delta I'_d = 6.2 - 2.9 = 3.3 \text{ mA}$
$\delta V_{gs} = -1 - (-2) = 1 \text{ V}$
$\delta I''_d = 3.0 - 2.9 = 0.1 \text{ mA}$
$\delta V_{ds} = 8 - 2 = 6 \text{ V}$

The mutual conductance g_m of the f.e.t. is given by the expression $g_m = \delta I_d / \delta V_{gs}$, with V_{ds} constant at 12 V. It can be seen from the characteristics that a change in V_{gs} from -2 V to -1 V produces a change in I_d from 2.9 to 6.2 mA. Therefore

$$g_m = \frac{6.2 - 2.9}{2 - 1} \times 10^{-3} = 3.3 \text{ mS} \qquad (Ans.)$$

Also from the characteristics it can be seen that a change in V_{ds} from 12 to 16 V, with V_{gs} constant at -2V, produces a change in I_d from 2.9 to 3.0 mA. Therefore

$$r_{ds} = \frac{16 - 12}{(3.0 - 2.9) \times 10^{-3}} = 40\,000 \ \Omega \qquad (Ans.)$$

The mutual characteristics of the junction f.e.t. are shown plotted in Fig. 1.19. The mutual conductance g_m of the device is given by the slope of the curve, thus for V_{ds} constant at 12 V.

$$g_m = \frac{(6.2 - 2.9) \times 10^{-3}}{1} = 3.3 \text{ mS} \qquad (Ans.)$$

Temperature Effects

The velocity with which majority charge carriers travel through the channel is dependent upon both the drain-source voltage and the temperature of the f.e.t. An increase in the temperature reduces the carrier velocity and this appears in the form of a reduction in the drain current which flows for given gate-source and drain-source voltages.

A further factor that may also affect the variation of drain current with change in temperature is the barrier potential

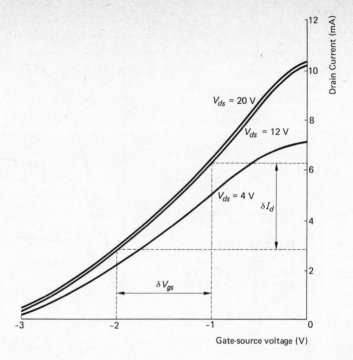

Fig. 1.19
$\delta V_{gs} = -1 - (-2) = 1$ V
$\delta I_d = 6.3 - 2.9 = 3.4$ mA

across the gate-channel p-n junction. An increase in temperature will cause the barrier potential to fall and this, in turn, will reduce the width of the depletion layer for a given gate-source voltage. The channel resistance will fall and the drain current will increase. The two effects tend to vary the drain current in opposite directions and as a result the overall variation can be quite small. Indeed, it is possible to choose a particular gate-source voltage and obtain zero temperature coefficient. In general, the overall result is that the drain current decreases with increase in temperature. This is the opposite of the collector current variation experienced by the bipolar transistor.

Handling the M.O.S.F.E.T.

The gate terminal of a m.o.s.f.e.t. is insulated from the channel by a very thin ($\simeq 100$ nm) layer of silicon dioxide, which effectively forms the dielectric of a capacitance. Any electric charge which accumulates on the gate terminal may easily produce a voltage across the dielectric that is of sufficient magnitude to break down the dielectric. Once this happens the gate is no longer insulated from the channel and the m.o.s.f.e.t. has been destroyed. The charge necessary to damage a m.o.s.f.e.t. need not be large since the capacitance between the gate and channel is very small and $V = Q/C$. This means that a dangerously high voltage can easily be produced by merely touching the gate leads with a finger or a tool. To prevent damage to m.o.s.f.e.t.s in store or about to be fitted into a circuit it is usual for them to be kept with their gate and source leads short-circuited together. The protective short-circuit can be provided by twisting the leads together, by means of a springy wire clip around the leads, or by inserting the leads into a conductive jelly or grease. The short-circuit must be retained in place while the device is fitted into a circuit, particularly during the soldering process.

Some m.o.s.f.e.t.s are manufactured with a zener diode internally connected between gate and substrate. Normally, the voltage across the diode is too low for it to operate and it has little effect on the operation of the device. If a large voltage should be developed at the gate by a static electric charge, the zener diode will break down before the voltage has risen to a value sufficiently great to cause damage.

The Field-Effect Transistor as a Switch

A field-effect transistor can be employed as an electronic switch since its drain current can be turned ON or OFF by the application of a suitable gate-source voltage. In the ON condition the gate-source voltage has moved the operating point to the top of the load line (see Fig. 1.20), and maximum drain current flows. The voltage across the f.e.t., known as the *saturation* voltage $V_{DS(SAT)}$, does not fall to zero but is typically in the range 0.2 V to 1.0 V. In the case of the f.e.t. characteristics illustrated, the ON resistance is 0.9 V/7.6 mA or

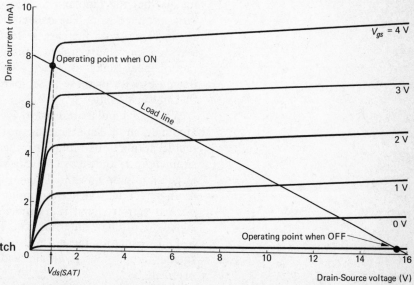

Fig. 1.20 The f.e.t. as a switch

118 Ω but this value is somewhat higher than many other f.e.t.s present; typically R_{ON} is some 30–60 Ω. To turn the f.e.t. OFF the gate-source voltage is reduced to, or below, the pinch-off figure, and the operating point is shifted to the bottom of the load line. The drain current is now reduced to a very small value, typically 1 nA for a junction f.e.t. and about 50 pA for a m o.s.f.e.t.

The minimum time taken by a f.e.t. to switch from one state to the other is another important feature. Power is mainly dissipated within a f.e.t. switch during the time it is passing from one state to the other since, when the device is ON or OFF, either the voltage across it or the current in it is very small, and power dissipation is the product of voltage and current. The faster the switching speed of a f.e.t. the higher its efficiency.

The Relative Merits of Bipolar Transistors and F.E.T.s

The input impedance of a bipolar transistor depends upon the d.c. collector current it conducts under quiescent conditions, and for the majority of applications it is somewhere in the region of 1000–3000 ohms. If the transistor is biased so that its quiescent collector current is only a few microamperes, an input impedance of a megohm or more can be achieved. The input impedance of a junction f.e.t. is very high, with a m.o.s.f.e.t. having an impedance which is several orders higher still. The mutual conductance of a bipolar transistor is 40 mS per mA of alternating collector current and is therefore considerably higher than the mutual conductance of a f.e.t.; this means that the bipolar transistor is capable of providing the larger voltage gain. The collector current of a bipolar transistor increases with increase in temperature and thermal runaway is a possibility unless suitable d.c. stabilization circuitry and/or heat sinks are used. The drain current of a f.e.t. decreases with increase in temperature and there is no risk of thermal instability.

When a bipolar transistor is used as a radio-frequency amplifier in a superheterodyne radio receiver, third-order (cubic) terms in its mutual characteristic will lead to an undesirable effect, known as CROSS-MODULATION, whenever a large-amplitude unwanted signal is applied. Cross-modulation is an effect in which the amplitude modulation on the unwanted carrier signal is transferred on to the wanted carrier signal to produce interference. The mutual characteristic of a f.e.t. contains these third-order terms at a very much smaller amplitude, if at all, and cross-modulation is not a problem.

When a f.e.t. is used as a switch its ON resistance is larger than the ON resistance obtainable from a transistor but the switching operation can be carried out in either direction, i.e. the drain and source terminals are interchangeable. On the other hand, the switching speed of the f.e.t. is slower than that of the bipolar transistor. This is because the ON resistance of a f.e.t. is larger than that of a bipolar transistor and so a f.e.t. is unable to charge or discharge the stray capacitances and the input capacitance of the next stage as quickly.

Exercises

1.1. The drain characteristics of a f.e.t. are given in Table A. Plot the characteristics and determine the drain-source resistance from the characteristic for $V_{gs} = 0.5$ V.

Use the curves to find the mutual conductance for $V_{ds} = 20$ V. What type of f.e.t. is this?

Table A

Drain current I_d (mA)

Drain-source voltage V_{ds}(V)	Gate-source voltage $V_{gs} = 1$ V	0.5 V	0 V	−0.5 V	−1 V
10	4.00	3.19	2.38	1.57	0.76
20	4.02	3.21	2.40	1.59	0.78
30	4.04	3.23	2.43	1.61	0.80

1.2. Draw sketches to show the various steps in the manufacture of an enhancement type m.o.s.f.e.t.

1.3. The data given in Table B refer to a f.e.t. Use the data to draw the drain characteristics, and then use the curves to determine: (i) the drain-source resistance for $V_{gs} = -3$ V, and (ii) the mutual conductance for $V_{ds} = -6$ V.

State the type of f.e.t. to which the data refer.

Table B

Drain current (mA)

Drain-source voltage $V_{ds} = $ (V)	Gate-source voltage $V_{gs} = -4$ V	−3 V	−2 V
−1	−6.0	−4.6	−3.1
−3	−6.6	−5.1	−3.5
−5	−7.2	−5.6	−3.9
−7	−7.8	−6.1	−4.3
−9	−8.4	−6.6	−4.7

1.4. An n-channel junction f.e.t. has the data given in Table C. Plot the drain and mutual characteristics and use them to determine the mutual conductance of the device. Find also the drain-source resistance.

Table C

Drain-source voltage V_{ds} (V)	Drain Current (mA)			
	Gate-source voltage $V_{gs} = 0$ V	−0.5 V	−1.0 V	−1.5 V
10	2.25	1.35	0.7	0.3
20	2.29	1.38	0.73	0.33
30	2.32	1.41	0.75	0.35

1.5. Explain, with the aid of a circuit diagram, how you would measure the drain and mutual characteristics of an n-channel m.o.s.f.e.t. Draw a typical set of drain characteristics and say how the characteristics of a p-channel m.o.s.f.e.t. would differ from those shown.

1.6. Describe, with the aid of sketches, the principles of operation of a p-channel junction field-effect transistor. Sketch a typical family of drain characteristics and account for their shape.

1.7. The drain characteristics of a field-effect transistor are as given in Fig. 1.21. Use the characteristics to obtain the values of drain current corresponding to various values of gate-source voltage when V_{ds} is (i) −3 V and (ii) −6 V and thence plot the mutual characteristics of the transistor for these drain-source voltages.

1.8. The relationship between the drain current and the drain-source voltage of a field effect transistor up to pinch-off is given by Table D.

Table D

V_{ds} (V)	0.5	1.0	1.5	2.0	2.5	3.0	3.5	4.0
I_d (mA)	0	5	7	8	11	12.5	14.7	16

Plot the characteristic over this range and thence determine the a.c. resistance of the transistor when (i) $V_{ds} = 1.0$ V, (ii) $V_{ds} = 2.0$ V, and (iii) $V_{ds} = 3.0$ V. Suggest a possible use of the device when operated on this part of its characteristics.

1.9. The output characteristics of a bipolar transistor are given in Table E. Plot the characteristics and use them to find the values of the h parameters, h_{OE}, h_{oe}, h_{FE} and h_{fe} when $V_{CE} = 12$ V.

Table E

Collector/emitter voltage V_{CE} (V)	Collector current $I_b = 20\ \mu A$	$80\ \mu A$	$140\ \mu A$
4	4.8	21.6	38.4
12	6.4	24.4	42.4
20	8.0	27.2	46.4

1.10. Describe the principle of operation, and give the symbol, of an enhancement-type m.o.s.f.e.t. Illustrate your answer with typical drain characteristics.

Short Exercises

1.11. A depletion-type m.o.s.f.e.t. can be used in either the enhancement or the depletion modes. Explain why this is not also true

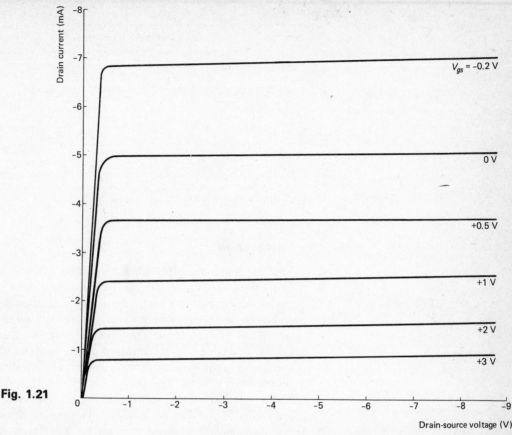

Fig. 1.21

for a junction f.e.t. or for an enhancement-type m.o.s.f.e.t. *NOT NORMAL DONE*

1.12. Compare the properties of a field-effect transistor with those of a bipolar transistor.

1.13. State the precautions that must be taken when using a m.o.s.f.e.t.

1.14. Draw the symbol for each type of field-effect transistor. How do the symbols show (i) the use of a p- or an n-channel, (ii) the use of an insulated gate, and (iii) whether or not a conducting path exists between drain and source when the gate-source voltage is zero?

1.15. Compare the relative merits of field-effect and bipolar transistors as electronic switches.

1.16. Why may an increase in temperature lead to thermal runaway in a bipolar transistor but not in a field-effect transistor?

1.17. (*a*) In which direction do majority charge carriers flow in an n-channel j.f.e.t.?

(*b*) A f.e.t. is operated as an amplifier with a positive drain supply voltage and its gate held positive relative to its source. What kind of f.e.t. is it?

(*c*) What is meant by pinch-off?

1.18. If the gate of an n-channel j.f.e.t. is taken positive with respect to the source by more than about 0.5 V the high input impedance feature of the device is lost. Explain the reason for this.

1.19. Sketch the construction of a p-channel j.f.e.t.

1.20. What is thermal runaway? Why is it a problem with bipolar transistors but not with f.e.t.s.?

2 Integrated Circuits

Introduction

The methods used to fabricate silicon planar bipolar and field-effect transistors can be extended to allow a complete circuit to be manufactured in a single silicon chip. All the components, active and passive, which are required by the circuit are formed at the same time in a small piece of silicon, known as a CHIP, by the diffused planar process. The circuit is known as a monolithic integrated circuit because only one silicon chip is used. The use of monolithic integrated circuits has a number of advantages over discrete circuits: greatly reduced size and weight, lower costs, complex circuit functions are economically possible, e.g. pocket calculators, and greater reliability. The size and weight reductions occur because a quite complex circuit can be enclosed within a volume of comparable dimensions to those of a single transistor. The cost of an integrated circuit depends upon its complexity and the quantity manufactured, but in many cases the cost is no greater than that of one transistor.

Other types of integrated circuit, known as thin-film and thick-film circuits are also available; with both thin-film and thick-film circuits, resistors and capacitors are fabricated by forming a suitable film onto the surface of a glass or a ceramic substrate. The components are interconnected in the required manner by means of a deposited metallic pattern. Thin film components are produced by vacuum deposition of a suitable material onto the surface of the substrate. Thick film components are produced by painting the substrate with special kinds of ink. Active components and inductors cannot be produced in this way and any such components that are necessary must be provided in discrete form, and be joined into the metallic pattern at the appropriate points. Thick- and thin-film circuits are not used to anywhere near the same extent as monolithic circuits and they will not be discussed in this book.

Integrated Circuit Components

The fabrication of an integrated circuit component is achieved by a sequential series of oxidizing, etching and diffusion, similar to that employed for the silicon planar bipolar transistor and the field-effect transistor (p. 9). The components that can be formed by this process are transistors, diodes, resistors and capacitors; inductors cannot be produced.

A thin wafer, about 5–10 mils† thick, is sliced from a rod of p-type silicon and will have a surface area of about 4 in². Since an integrated circuit may only occupy an area of about 30 mils², several thousands of identical circuits can be simultaneously fabricated in the one wafer. The principle is illustrated by Fig. 2.1, although to simplify the drawing fewer circuits have been shown.

Each individual silicon chip acts as a substrate into which the various components making up the circuit can be formed. The components are simultaneously formed by the diffusion of impurity elements into selected parts of the chip.

Since the p-type silicon substrate is an electrical conductor it is necessary to arrange that each of the components is insulated from the substrate. If this is not done the various components will all be coupled together by the substrate resistance. There are a number of different ways in which the required isolation can be obtained, but the most common method utilizes the high-resistance property of a reverse-biased p-n junction (see Fig. 2.2). Several n-type regions, equal in number to the number of components in the circuit, are diffused into the p-type substrate. Each of the n-type regions will be isolated from the substrate if the junction is maintained in the reverse-bias condition by connecting the substrate to a potential which is more negative than any other part of the circuit.

The various components making up the circuit are fabricated by means of a number of n-type and p-type regions which are diffused into the isolated regions. Once formed, the components are interconnected as required by the circuit by means of an aluminium pattern deposited onto the surface of the chip.

Integrated Bipolar Transistor

The most commonly used active device in an integrated circuit is the n-p-n bipolar transistor, the construction of which is shown in Fig. 2.3. (n^+ denotes a region of greater conductivity.) The construction is similar to that of the silicon planar

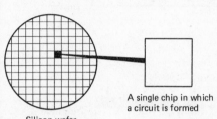

Silicon wafer

Fig. 2.1 Showing how a silicon wafer is divided into a number of chips

A single chip in which a circuit is formed

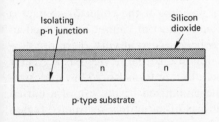

Fig. 2.2 Method of isolating the components in an integrated circuit

† One mil is one thousandth of an inch.

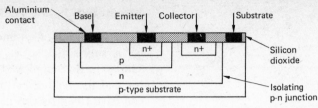

Fig. 2.3 An integrated n-p-n bipolar transistor

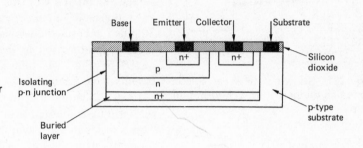

Fig. 2.4 An integrated n-p-n bipolar transistor with a buried layer

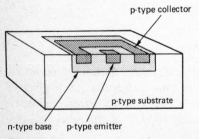

Fig. 2.5 Lateral p-n-p bipolar transistor

transistor [E II], but differs from it in that the collector contact is brought out at the top of the transistor instead of at the bottom. The change in the position of the collector contact is necessary because the collector current cannot be allowed to flow in the substrate. The collector current must therefore flow in the narrow collector region and so the device has a greater collector resistance than the discrete transistor. This, undesirable, series resistance can be minimized by the use of a buried layer. The buried layer consists of an n^+ low-resistance region diffused into the chip in the position shown in Fig. 2.4. The buried layer is effectively in parallel with the collector region and reduces the collector series resistance. The series resistance cannot be reduced by using a lower resistivity collector region since this would reduce the breakdown voltage of the collector-base junction. Typically, an h_{fe} value of about 100 is achieved. At high frequencies, the capacitance of the isolating p-n junction may possess a sufficiently low reactance to couple the collector to the substrate and adversely affect the frequency response.

The fabrication of a p-n-p transistor is not as simple or as cheap because additional p-type and n-type regions are required. Alternatively, a different and less efficient layout known as a *lateral transistor* can be employed which is more expensive and provides a lower current gain of about 5. Because of the difficulties associated with the use of the p-n-p transistor, its use in an integrated circuit is avoided whenever possible. (See Fig. 2.5.)

Integrated M.O.S.F.E.T.

The constructional details of a m.o.s.f.e.t. in an integrated circuit are the same as those of a discrete m.o.s.f.e.t. (see Chapter 1). The m.o.s.f.e.t. has an advantage over the bipolar transistor in that it is self-isolating; the drain and source regions are each isolated from the substrate by their individual p-n junctions, while the gate terminal is isolated by a layer of silicon dioxide. This feature allows a m.o.s.f.e.t. to be formed in a smaller area of the chip than can be achieved with a bipolar transistor. Because of this, a m.o.s.f.e.t. is sometimes used with its gate and drain terminals connected together, when it acts as a resistor.

Integrated Diode

Fig. 2.6 shows the construction of an integrated circuit diode. The diode is formed at the same time as one of the junctions of a transistor and consists of a p-type region (the anode) and an n-type region (the cathode). An n^+ region is diffused into the chip to reduce the resistance of the cathode contact.

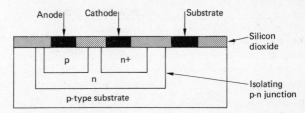

Fig. 2.6 An integrated diode

Integrated Resistor

Integrated resistors are made using a thin layer of p-type silicon that is diffused at the same time as the base of the transistor. The resistance of a silicon layer depends upon the length l, area a and resistivity ρ, of the layer according to equation (2.1), i.e.

$$R = \frac{\rho l}{a} \tag{2.1}$$

The area a of the layer is the product of width W and the depth, d of the layer. Thus

$$R = \rho l / W d \ \Omega$$

It is usual to express the resistance in terms of the resistance of a square of the silicon layer (Fig. 2.7) in which the width W of the layer is equal to the length l. Then, equation (2.1) can be written as

$$R = \frac{\rho l}{ld} = \frac{\rho}{d} \ \Omega/\square \tag{2.2}$$

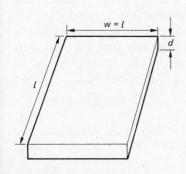

Fig. 2.7

The resistance is now the resistance between the opposite sides of a square and it is measured in a unit known as the ohm per square. The resistance depends only upon the resistivity of the silicon layer and not upon the dimensions of the square. The resistivity of the layer is determined by the number of charge carriers (holes) that are diffused into the layer and the depth to which they penetrate. However, since both of these variables are fixed by the requirements of the simultaneously diffused transistors, a required resistance value must be obtained by a suitable choice of the length and width of the resistive path. The resistance value given by a square can be increased by increasing the length of the path, or decreased by increasing the width of the path. Difficulties are experienced with the fabrication of very high values of resistance because of the relatively large chip area such resistances demand.

The constructional details of an integrated resistor are given in Figs. 2.8a and b. Fig. 2.8a shows that the resistive path is

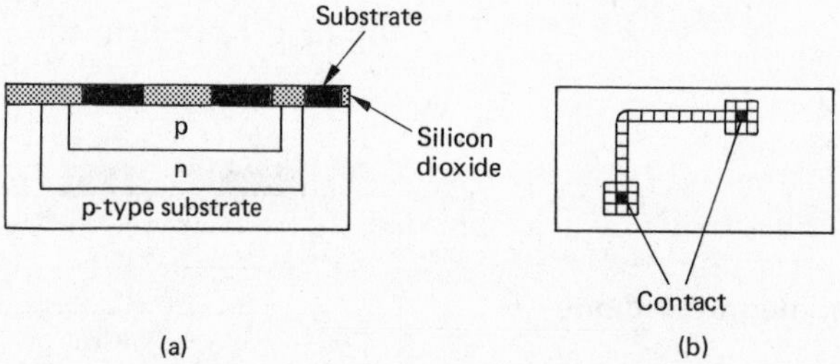

Fig. 2.8 Side and top views of an integrated resistor

(a)　　　　　　(b)

formed by a p-type region that joins together the resistor contacts. A p-type path is used since it will be diffused at the same time as transistor base regions and will therefore be only lightly doped. The resistivity will therefore be in the range of 100–300 $\Omega/\square$. When very low values of resistance are required an n-type resistor is employed; the n-type path is diffused at the same time as the transistor emitter regions and will therefore be of much lower resistivity. Fig. 2.8b shows the top view of an integrated resistor and indicates how a required resistance value may be obtained by connecting in series a number of "squares". The resistor can follow any path that will best utilize the surface area of the chip. The practical range of resistance values is from about 15 Ω to about 30 kΩ.

EXAMPLE 2.1

The resistivity of a p-type region is 100 $\Omega/\square$. Calculate the resistance of a strip which is 1 mil wide and (i) 20 mils long, (ii) 30 mils long.

Solution
Since the resistive strip is 1 mil wide it will have a resistance of 100 Ω per 1 mil length.

(i) The resistance of a 20 mil length is $20 \times 100 = 2000 \, \Omega$
(ii) The resistance of a 30 mil length is $30 \times 100 = 3000 \, \Omega$

Integrated Capacitor

Integrated capacitors can be fabricated in two ways: either the capacitance of a reverse-biased p-n junction can be utilized, or the capacitance can be provided by a layer of silicon dioxide separating two conducting areas. The construction of a junction-type capacitor is shown in Fig. 2.9a. The p-n junction is formed at the same time as either the emitter-base or the collector-base junction of a transistor. Provided the p-n junction is held in the reverse-biased condition, a capacitance of about 0.2 pF/mil can be obtained. Since the area of the chip available for a capacitor is limited, values of up to about 100 pF are available. Fig. 2.9b shows a m.o.s. capacitor; one

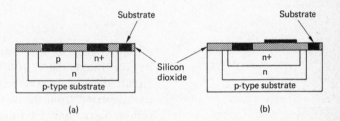

Fig. 2.9 (a) Integrated capacitor and (b) integrated m.o.s. capacitor

electrode of the capacitor is provided by an aluminium layer that is deposited onto the top of the silicon layer and the other electrode is produced by the diffused n^+ region. The capacitance provided depends upon the thickness of the silicon dioxide layer and the area of the aluminium plate; up to a few hundred picofarads can be achieved.

The m.o.s. capacitor is more expensive to provide but has the following advantages over the junction capacitor: it can have voltages of either polarity applied to it, it has lower losses and a larger breakdown voltage, and its capacitance value does not depend upon the magnitude of the voltage applied across the capacitor.

The Fabrication of a Complete Integrated Circuit

A linear integrated circuit is one in which the output signal is proportional to the input signal. Many linear integrated circuits are amplifiers of one kind or another, while many others perform functions that are beyond the scope of this book.

The main differences between integrated and linear circuits which perform the same function are that the i.c. uses n-p-n transistors and diodes as liberally as possible. This is because resistors and capacitors occupy more space in the chip than transistors and are therefore more expensive.

In the fabrication of a complete integrated circuit all the components, active and passive, required to make up the circuit are formed at the same time. The components are then interconnected as required by means of an aluminium pattern which is deposited on the top of the silicon slice. As an

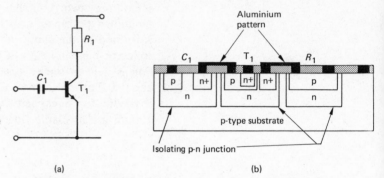

Fig. 2.10 Showing (a) a simple transistor circuit and (b) the same circuit in integrated form

example, suppose the simple circuit shown in Fig. 2.10a is to be integrated. Fig. 2.10b shows the three components of the circuit diffused into a p-type substrate. The components are each isolated from the substrate by a reverse-biased p-n junction and are connected together in the required manner by an aluminium pattern which is deposited onto the surface of the chip.

A large number of circuits are simultaneously produced in a single silicon wafer, and after formation they are separated into individual chips and then sealed within a suitable package. The majority of integrated circuits are available in one or more of two kinds of package; these packages are the TO circular packages and the dual-in-line. The two packages are illustrated by Figs. 2.11a and b, the latter being perhaps the most popular.

Practical Circuits

A wide variety of linear integrated circuits are available and include many different kinds of amplifier. The main types of amplifier available are: (i) audio-frequency amplifiers, (ii) wideband amplifiers, (iii) radio-frequency amplifiers, and (iv) operational amplifiers.

The early monolithic integrated circuits were all digital in nature, partly because digital circuitry is easier to fabricate but also because the potential demand for similar digital circuits was very much larger. The first linear monolithic integrated circuits were operational amplifiers (discussed in Chapter 4)

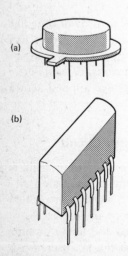

Fig. 2.11

which can be used for a wide variety of applications. The early operational amplifiers were prone to self-oscillation and latching but this problem has been overcome in the modern amplifiers. Audio amplifiers are available with output powers of up to 10 watts with built-in protection against damage caused by overload and output short-circuits. A variety of radio-frequency amplifier integrated circuits are offered by different manufacturers. Most of these circuits incorporate an automatic gain control facility and many also include mixing and/or detector circuits as well.

When an integrated circuit appears in an equipment, a number of additional external components are also required. Operational amplifiers must have external resistors connected between the appropriate terminals to specify the overall voltage gain of the circuit, and sometimes one or more external capacitors are needed to ensure stability. Decoupling capacitors when needed must also be provided externally since the values required are too large to be fabricated in the silicon chip. The power supplies to an operational amplifier are normally decoupled while audio amplifiers require emitter decoupling capacitors. Variable resistors, such as volume and tone controls, must also be externally provided. Inductors cannot be produced within a monolithic integrated circuit and hence the desired selectivity of a radio-frequency amplifier must be specified by external components. Usually, a parallel-resonant tuned circuit is connected across the appropriate terminals for this purpose, but some circuits utilize the selectivity characteristics of ceramic or crystal filters.

Exercises

2.1. (a) Describe, using appropriate sketches, how either (i) an enhancement type m.o.s.f.e.t. or (ii) a bipolar transistor can be fabricated in integrated circuit form.

(b) Briefly explain why high value resistors and capacitors are not included in monolithic integrated circuitry.   (C&G)

2.2. (a) Explain, with appropriate sketches, how: (i) a capacitor, (ii) a resistor, (iii) a diode can be formed on a silicon slice in the manufacture of an integrated circuit.

(b) Describe any method which can be used to isolate electrically each component on the wafer.   (C&G)

2.3. Describe with the aid of a diagram the construction of a low-value (about 20 Ω) integrated resistor.

2.4. The resistive path of an integrated resistor has a resistivity of 250 $\Omega/\square$. Calculate the length of path required to produce a resistance of 3750 Ω if the width of the path is: (a) 1 mil, and (b) 2 mil.

2.5. Explain, with sketches where necessary, the meaning of four of the following when applied to solid state device manufacture: (i) monolithic integrated circuit, (ii) planar process, (iii) diffusion, (iv) epitaxial layer, (v) metallization.   (C&G)

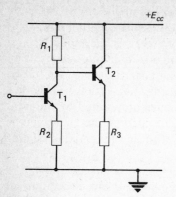

Fig. 2.12

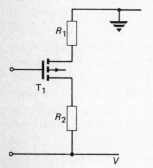

Fig. 2.13

2.6. (*a*) Describe, with the aid of sketches, the step-by-step fabrication of an integrated circuit field-effect transistor (MOST).

(*b*) Name two advantages gained by forming a MOST in preference to a bipolar transistor in an integrated circuit chip.

(C&G)

2.7. The circuit shown in Fig. 2.12 is to be integrated. Draw side and plan views of a possible arrangement of the integrated circuit.

2.8. (*a*) Explain how the various elements in a monolithic integrated circuit may be isolated from one another.

(*b*) In order to reduce the $V_{CE(SAT)}$ of a transistor in an integrated circuit, buried n^+ regions are often incorporated. With the aid of a diagram explain the principle of this technique.

(*c*) Describe briefly why in many cases an integrated bipolar transistor has a poorer h.f. response than its discrete counterpart.

(C&G)

2.9. Draw a sketch to show the integrated circuit version of the circuit shown in Fig. 2.13. What should be the polarity of the voltage *V*?

2.10. Fig. 2.14 shows a diagram of an integrated circuit which is intended for use in a medium and long waveband type of radio receiver. Pins 5 and 13 are power supply connections, pin 4 is a bias supply, and pins 6 and 14 are chassis connections. List briefly, giving reasons, the external circuit components to which the remaining pins would be connected. (part C&G)

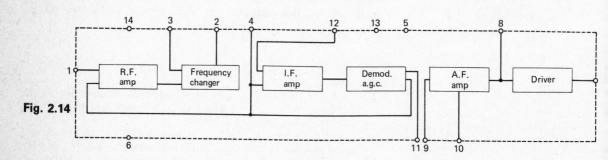

Fig. 2.14

Short Exercises

2.11. What is meant by each of the following terms used in conjunction with integrated circuits: (i) buried layer, (ii) substrate, and (iii) mil?

2.12. Why must high values of resistance and capacitance be provided by components external to a monolithic integrated circuit?

2.13. What is the main difference between a discrete bipolar transistor and an integrated bipolar transistor? What are the advantages of an integrated m.o.s.f.e.t. over an integrated bipolar transistor?

2.14. Draw the construction of an integrated m.o.s.f.e.t. State which type you have drawn.

2.15. What is the difference between a junction capacitor and a m.o.s. capacitor in an integrated circuit? Give values for the maximum capacitance value of each type.

2.16. Why is it necessary to isolate an integrated bipolar transistor from the substrate of its chip? Why is it not necessary to provide such isolation for an integrated m.o.s.f.e.t.?

2.17. An integrated resistor has a resistance of 500 Ω. What will be the resistance value if (i) the length of the resistive path was doubled, (ii) the width of the resistive path was doubled?

2.18. List the advantages of monolithic integrated circuits.

3 Small-signal Audio-frequency Amplifiers

Principles of Operation

The active device used in an amplifier may be a bipolar transistor or a field-effect transistor. For brevity, in the remainder of this book the bipolar transistor will be referred to as a *transistor*, and the field-effect transistor as a *f.e.t.*

Transistors and f.e.t.s may be used as amplifiers because their output currents can be controlled by an a.c. signal applied to their input terminals. If a voltage or power output is required the output current must be passed through a resistive load. A f.e.t. has such a high input impedance that its input current is negligible; it can therefore give only a voltage gain. The input impedance of a transistor depends upon the magnitude of the collector current it is passing and can be made fairly large if the circuit is biased so that the collector current is only a few microamperes. By suitable choice of collector current, and hence of input impedance, a transistor may be considered as either a current- or a voltage-operated device. If the source impedance is much larger than the input impedance of the transistor, the transistor is current operated; if much smaller, it is voltage operated. For reasons given later, an audio-frequency transistor is usually operated in the common-emitter connection, when its short-circuit a.c. current gain is $h_{fe} = \delta I_c / \delta I_b$. The collector current is then equal to $h_{fe} \delta I_b$. Alternatively, the a.c. collector current may be expressed in terms of the mutual conductance, g_m, of the transistor:

$$g_m = \frac{\delta I_c}{\delta V_{be}} \qquad (V_{ce} \quad \text{constant}) \qquad (3.1)$$

This equation is valid over a wide range of collector currents (say $0.1\,\mu\text{A}$ to $2\,\text{mA}$), and over a frequency range from zero (direct current) to about $1\,\text{MHz}$. Dividing h_{fe} by g_m,

$$\frac{h_{fe}}{g_m} = \frac{\delta I_c}{\delta I_b}\frac{\delta V_{be}}{\delta I_c} = \frac{\delta V_{be}}{\delta I_b}$$

$\delta V_{be}/\delta I_b$ is the ratio of the voltage applied across the input terminals to the current flowing into the transistor, i.e. the input impedance. From this equation,

$$\text{Short-circuit current gain} = \text{Mutual conductance} \times \text{Input impedance} \tag{3.2}$$

Voltage operation of a transistor provides the following advantages: (a) the 3 dB bandwidth of g_m is considerably greater than that of h_{fe}, (b) less noise is introduced, and (c) g_m is primarily determined by the magnitude of the a.c. component of the collector current and is more predictable than h_{fe}. On the other hand, current operation gives a greater stage gain, and because a larger collector current is employed, a larger maximum permissible output signal. Voltage operation is generally restricted to the input stage of an amplifier.

The dynamic mutual characteristics of a f.e.t. or a transistor always exhibit some non-linearity. If a suitable operating point is chosen and the amplitude of the input signal is limited, the operation of the circuit may be taken as linear without the introduction of undue error.

The function of a *small-signal amplifier* is to supply a current or voltage to a load, the power output being unimportant. In a *large-signal amplifier*, on the other hand, the power output is the important factor and to obtain an adequate power output the output current and voltage swings must cover most of the characteristics.

It was shown in [EII] that the current, voltage and power gains of a transistor amplifier are given by

$$\text{Current gain } A_i = \frac{\text{Change in output current}}{\text{Change in input current}} \tag{3.3}$$

$$\text{Voltage gain } A_v = \frac{\text{Change in output voltage}}{\text{Change in input voltage}} = \frac{A_i R_L}{R_{IN}} \tag{3.4}$$

$$\text{Power gain } A_p = \frac{\text{Change in output power}}{\text{Change in input power}} = \frac{A_i^2 R_L}{R_{IN}} = A_i A_v \tag{3.5}$$

Choice of Configuration

The various ways in which a transistor or f.e.t. may be connected to provide a gain are shown in Fig. 3.1 (npn transistors and n-channel f.e.t.s are shown).

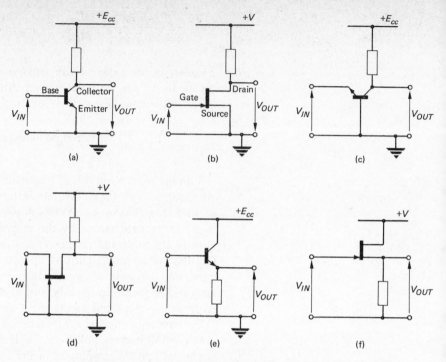

Fig. 3.1 Possible connections for transistors and f.e.t.s

A transistor connected as a common-base amplifier (Fig. 3.1c) has a short-circuit a.c. current gain h_{fb}, less than unity (typically about 0.992), a low input impedance of the order of 50 Ω, and an output impedance of about 1 MΩ. Because the current gain is less than unity, common-base stages cannot be cascaded using resistance-capacitance coupling but transformer coupling can be used. Transformers, however, have the disadvantages of being relatively costly, bulky and heavy and having a limited frequency response, particularly the miniature types used in transistor circuits.

The short-circuit a.c. current gain h_{fe} of a transistor connected in the common-emitter configuration (Fig. 3.1a) is much greater than the short-circuit a.c. current gain of the same transistor connected with common base, i.e. $h_{fe} = h_{fb}/(1 - h_{fb})$. Resistance-capacitance coupling of the cascaded stages of an amplifier is possible and nowadays transformers are rarely used. Generally, common-emitter stages are biased so that the transistor is current operated. Then the input impedance is in the region of 1000–2000 Ω while the output impedance is some 10–30 kΩ.

The common-collector circuit, or EMITTER FOLLOWER as it is usually called, is shown in Fig. 3.1e. This connection has a high input impedance, a low output impedance, and a voltage gain less than unity. The main use of an emitter follower is as a power-amplifying device that can be conveniently connected between a high-impedance source and a low-impedance load.

The three terminals of an f.e.t. are called the *source, gate* and *drain*, corresponding to the emitter, base and collector of a transistor. In the normal mode of operation (Fig. 3.1*b*) the source is common to the input and output circuits, the input signal is applied to the gate, and the output is taken from between drain and earth. This connection provides a large voltage gain and has a high input impedance.

Fig. 3.1*f* shows the f.e.t. equivalent of the emitter follower; this is known as the SOURCE FOLLOWER circuit. The follower circuit will be treated in greater detail in Chapter 4. Lastly, Fig. 3.1*d* shows the common gate connection; this is not used at audio frequencies.

Determination of Gain using a Load Line

The *short-circuit* current gain of a transistor connected in the common-emitter configuration is $h_{fe} = \delta I_c / \delta I_b$ and the gain of a f.e.t. is best described by its mutual conductance g_m. When a load resistance is connected in the collector or drain circuit of the transistor or f.e.t., the available gain falls by a factor that depends upon the load resistance. The current or voltage gain of a resistance-loaded amplifier can be found with the aid of a *load line* drawn on the output or drain characteristics of the device. Fig. 3.2 shows the current and voltage relationships of single-stage resistance-loaded amplifiers. The collector or drain current flows in the load resistance R_L and develops a voltage across it. The direct voltage applied across the transistor or f.e.t. is the supply voltage minus this voltage drop.

Applying Kirchhoff's second law to the circuits shown gives the following expressions:

$$E_{cc} = V_{ce} + I_c R_L \qquad (3.6a)$$

$$V = V_{ds} + I_d R_L \qquad (3.6b)$$

These equations are of the form $y = mx + c$ and are therefore equations to straight lines; to draw the line representing each equation on the output or drain characteristics it is only necessary to determine two points. These two points can best be found in the following manner.

(a) Let $I_c = I_d = 0$; then $E_{cc} = V_{ce}$ and $V = V_{ds}$, giving one of the required points.

(b) Let $V_{ce} = V_{ds} = 0$; then

$$E_{cc} = I_c R_L \quad \text{or} \quad I_c = E_{cc}/R_L$$

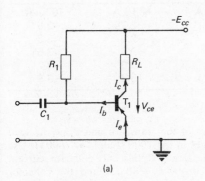

(a)

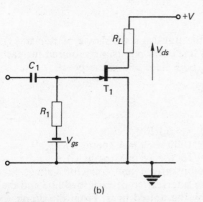

(b)

Fig. 3.2 Currents and voltages in basic amplifiers

and

$$V = I_d R_L \quad \text{or} \quad I_d = V/R_L$$

giving the second point.

If these two points are located on the characteristics and joined by a straight line, the *load line* for the particular load resistance and supply voltage is obtained. The load line can be used to determine the currents and voltages in the output circuit, since the ordinate of the point of intersection of the load line and a given input current or voltage curve gives the output current and voltage for that input signal.

EXAMPLE 3.1

A transistor connected in the common-emitter configuration has the data given in Table 3.1.

Table 3.1

	I_c (mA)		
V_{ce} (V)	$I_b = -60\ \mu A$	$-80\ \mu A$	$-100\ \mu A$
−1	−1.7	−2.25	−2.9
−3	−1.95	−2.6	−3.3
−5	−2.1	−2.8	−3.5
−7	−2.25	−3.0	−3.7
−9	−2.4	−3.2	−3.9

Plot the output characteristics of the transistor and draw the load lines for collector load resistances of (a) 1000 Ω and (b) 1800 Ω. Use the load lines to determine the steady (quiescent) collector current and voltage if the base bias current is −80 μA and the collector supply voltage is −9 V.

Solution
The output characteristics of the transistor are shown plotted in Fig. 3.3. To plot the required load lines, two points are required for each line. One point is common to both lines and is the point

$$I_c = 0 \qquad V_{ce} = E_{cc} = -9\ \text{V}$$

and is marked A in Fig. 3.3.
    The other point is $V_{ce} = 0$, $I_c = -E_{cc}/R_L$.
    For the 1000 Ω load $I_c = -9/1000 = -9\ mA$ (point B).
    For the 1800 Ω load $I_c = -9/1800 = -5\ \text{mA}$ (point C).
    The steady values of collector current and collector voltage are obtained by projecting from the intersection of each load line and the $I_b = -80\ \mu A$ curve as shown by the dotted lines. From the graph,
        when $R_L = 1000\ \Omega$, $I_c = -2.84\ \text{mA}$, $V_{ce} = -5.16\ \text{V}$
        when $R_L = 1800\ \Omega$, $I_c = -2.6\ \text{mA}$, $V_{ce} = -3\ \text{V}$.

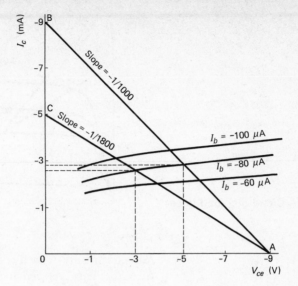

Fig. 3.3

Choice of Operating Point

The steady (quiescent) values of collector or drain voltage and current are fixed by the bias current or voltage. The application of an a.c. signal to the input terminals of the device causes the voltages and currents to vary about their quiesent values. Ideally, the input and output signal waveforms should be identical; in practice, however, some non-linearity always exists, and to minimize signal distortion care must be taken to restrict operation to the most nearly linear part of the characteristic. For this a suitable operating point must be selected and the signal amplitude must be restricted. The chosen operating point is fixed by the application of a steady bias current or voltage, using one of the bias arrangements discussed later in this chapter. The transistor is said to operate under Class A conditions.

A.C. Load Lines

Very often the load into which the transistor or f.e.t. works is not the same for both a.c. and d.c. conditions. When this is the case two load lines must be drawn on the characteristics: a d.c. load line to determine the operating point, and an a.c. load line to determine the current or voltage gain of the circuit. The a.c. load line *must* pass through the operating point.

Fig. 3.4 shows the circuit of a single-stage common-emitter amplifier using potential-divider bias. The emitter decoupling capacitor C_3 has a very high reactance at very low frequencies and does not shunt the emitter resistance R_4 at zero frequency

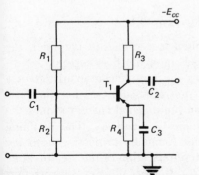

Fig. 3.4 A single-stage common-emitter amplifier

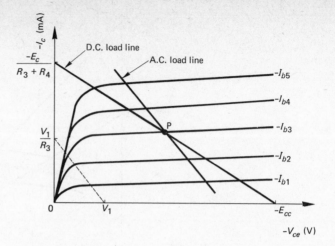

Fig. 3.5 A.C and d.c. load lines

(direct current). The d.c. load on the transistor is therefore $R_3 + R_4$ ohms. At signal frequencies the reactance of C_3 is low and the a.c. load on the transistor is reduced to R_3 ohms. A d.c. load line is first drawn on the output characteristics between the points

$$I_c = 0, \qquad V_{ce} = -E_{cc}, \quad \text{and} \quad V_{ce} = 0,$$
$$\text{and} \quad V_{ce} = 0, \qquad I_c = -E_{cc}/(R_3 + R_4)$$

(see Fig. 3.5). Suitable operating point P is then selected.

The a.c. load line must be drawn passing through the operating point with a slope equal to the reciprocal of the a.c. load on the transistor, i.e. $-1/R_3$. To avoid extending the current axis, proceed as follows: (*a*) Assume that the a.c. load is actually a d.c. load, and using any convenient value of supply voltage (V_1 in Fig. 3.5), draw lightly the corresponding d.c. load line using the method previously explained, i.e. between points V_1 and V_1/R_3. (*b*) This load line has the required slope, so draw the actual a.c. load line parallel to it and passing through the operating point.

Current Gain of a Transistor Amplifier

When an input signal is applied to a transistor amplifier, the signal current is superimposed upon the bias current. Referring to Fig. 3.6, which shows the output characteristics of a common-emitter connected transistor, suppose that the base bias current is I_{b2} and that an input signal current swings the base current between the values I_{b1} and I_{b3}. The resulting values of collector current are found by projecting onto the collector-current axis from the intersection of the a.c. load line and the curves for I_{b1} and I_{b3}. If the collector currents thus found are $I_{c(max)}$ and $I_{c(min)}$ then the current gain of the circuit is

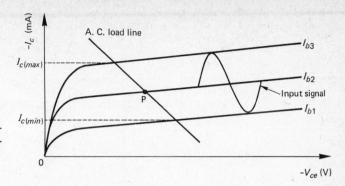

Fig. 3.6 Use of a.c. load line to calculate the current gain of a transistor amplifier

$$A_i = \frac{\text{Peak-to-peak collector current}}{\text{Peak-to-peak base current}} \qquad (3.7a)$$

$$= \frac{I_{c(max)} - I_{c(min)}}{I_{b3} - I_{b1}} \qquad (3.7b)$$

The peak-peak values are taken because two half-cycles of a distorted waveform may have different peak values.

EXAMPLE 3.2

The transistor used in the circuit of Fig. 3.7 has the data given in Table 3.2. Plot the output characteristics of the transistor, draw the d.c. load line and select a suitable operating point. Draw the a.c. load line and use it to find the alternating current that flows in the 2500 Ω load when an input signal producing a base current swing of $\pm30~\mu A$ about the bias current is applied to the circuit. Assume all the capacitors have zero reactance at signal frequencies.

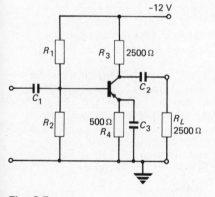

Fig. 3.7

Table 3.2

	I_c (mA)			
V_{ce} (V)	$I_b = -20~\mu A$	$I_b = -40~\mu A$	$I_b = -60~\mu A$	$I_b = -80~\mu A$
-3	-0.85	-1.55	-2.32	-3.08
-5	-1.00	-1.74	-2.56	-3.35
-7	-1.13	-1.92	-2.76	-3.60
-9	-1.30	-2.13	-3.00	

Solution
The output characteristics are shown plotted in Fig. 3.8. The d.c. load on the transistor is $R_3 + R_4 = 3000~\Omega$; the d.c. load line must therefore be drawn between the points

$$I_c = 0, \qquad V_{ce} = -12~V$$

and

$$V_{ce} = 0, \qquad I_c = -12/3000 = -4~mA.$$

Since the input signal has a peak value of $\pm30~\mu A$ a suitable base bias current is $-50~\mu A$; the operating point is then P.

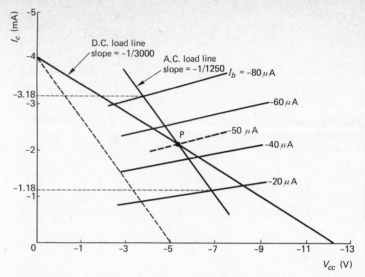

Fig. 3.8

The a.c. load on the transistor is the 2500 Ω collector resistor in parallel with the 2500 Ω load, i.e. 1250 Ω. An a.c. load line with a slope of −1/1250 must therefore be drawn. To draw a d.c. load line with the same slope assume a convenient supply voltage, say −5 V; then this d.c. load line joins the points

$$V_{ce} = -5 \text{ V} \quad \text{and} \quad I_c = -5/1250 = 4 \ mA$$

The equivalent d.c. load line is shown dotted and the wanted a.c. load line has been drawn parallel to it and passing through the operating point.

When a signal of ±30 μA (peak) is applied to the transistor the base current varies between −80 μA and −20 μA. Projection from the intersection of the load line and the −80 μA and −20 μA base current curves gives the resulting values of collector current as −3.18 mA and −1.18 mA. The peak collector signal current is therefore (3.18 − 1.18)/2, or 1.0 mA. The collector and load resistors are in parallel at signal frequencies and are both 2500 Ω, so the collector signal current divides equally between them. Therefore

$$\text{Load current} = 0.5 \text{ mA} \quad (Ans.)$$

Voltage Gain of F.E.T. Amplifier

The voltage gain of a f.e.t. can also be found with the aid of a load line. For example, Fig. 3.9 shows an a.c. load line drawn on the drain characteristics of a f.e.t., and the dotted projections from the load line show how the drain voltage swing, resulting from the application of an input signal voltage, can be found. The voltage gain A_v of the f.e.t. amplifier stage is

$$A_v = \frac{\text{Peak-to-peak drain voltage}}{\text{Peak-to-peak gate-source voltage}} \quad (3.8a)$$

$$= \frac{V_{ds(max)} - V_{ds(min)}}{V_{gs3} - V_{gs1}} \quad (3.8b)$$

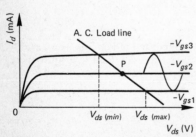

Fig. 3.9 Use of a.c. load line to calculate the voltage gain of a f.e.t. amplifier

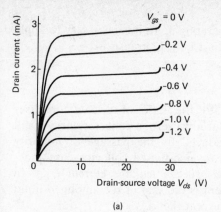

(a)

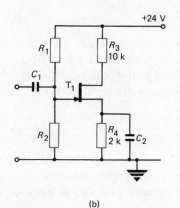

(b)

Fig. 3.10

EXAMPLE 3.3

Fig. 3.10a shows the drain characteristics of a common-source n-channel junction f.e.t., which is used in the single-stage amplifier circuit shown at (b). Draw the d.c. load line and select a suitable operating point. Draw the a.c. load line and use it to find the voltage gain when a sinusoidal input signal of 0.3 V peak is applied.

Solution

The d.c. load is 12 kΩ, and hence the d.c. load line must join the points

$$I_d = 0, \qquad V_{ds} = 24 \text{ V}$$

and

$$V_{ds} = 0, \qquad I_d = V/(R_3 + R_4) = 24/(12 \times 10^3) = 2 \text{ mA}$$

(see Fig. 3.11). A suitable operating point is $V_{gs} = -0.9$ V. The a.c. load line must pass through the chosen operating point with a slope of $-1/(10 \times 10^3)$ and has been drawn parallel to the dotted line joining the points

$$I_d = 0, \ V_{ds} = 24 \text{ V} \quad \text{and} \quad V_{ds} = 0, \ I_d = 24/10 \times 10^3 = 2.4 \text{ mA}$$

From the a.c. load line, the voltage gain of the circuit is

$$A_v = \frac{17 - 7}{-1.2 - (-0.6)} = -16.7 \qquad (Ans.)$$

Fig. 3.11

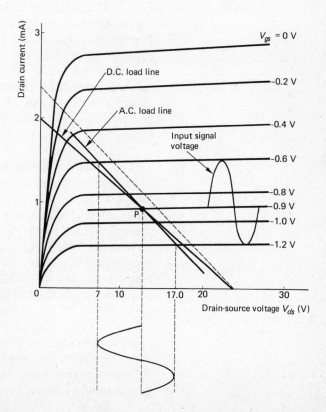

Bias and Stabilization

To establish the chosen operating point it is necessary to apply a bias voltage or current to a f.e.t. or transistor.

Transistor Bias

If the current flowing into the base of a common-base-connected transistor is reduced to zero, a collector current still flows. This current is produced by the passage of minority charge-carriers across the reverse-biased collector-base junction and is known as the *collector leakage current* I_{CBO}. In general, the collector current is the sum of the amplified input current and the leakage current, i.e.

$$I_c = h_{FB}I_e + I_{CBO} \tag{3.9}$$

I_{CBO} may be only 10 nA for a silicon planar transistor and perhaps a few microamperes for a germanium transistor.

When a transistor is connected in the common-emitter configuration the base current becomes the input current and the current gain is increased. Then

$$I_c = h_{FE}I_b + I_{CEO} \tag{3.10}$$

I_{CEO} is the collector leakage current of a common-emitter-connected transistor; I_{CEO} is considerably larger than I_{CBO}. The quantity h_{FE} is the d.c. current gain and is very nearly equal to the a.c. current gain h_{fe}.

An increase in the temperature of the collector–base junction will produce an increase in I_{CBO}. The resulting increase in collector current gives an increase in the power dissipated at the junction, and this, in turn, increases still further the temperature of the junction and gives a further increase in I_{CBO}. The process is cumulative, and particularly in the common-emitter connection (since $I_{CEO} \gg I_{CBO}$), leads to signal distortion caused by the operating point moving along the load line. In extreme cases the eventual destruction of the transistor may occur; this is known as *thermal runaway*. To prevent thermal runaway it is often necessary to employ a bias current that also gives some degree of d.c. stabilization. The current gain h_{FE} and base–emitter voltage are also functions of temperature and can lead to changes in collector current. In addition, individual transistors of a given nominal h_{FE} may have values of h_{FE} lying between quoted maximum and minimum values. For example, one manufacturer offers transistors with h_{FE} in the ranges 20–125 and 125–500.

An amplifier stage will be designed to have a particular d.c. collector current using the nominal value of h_{FE}. The bias circuit should operate to ensure that approximately the same

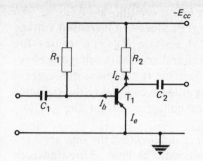

Fig. 3.12 Fixed bias

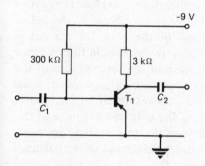

Fig. 3.13

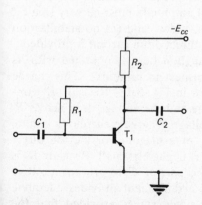

Fig. 3.14 Collector-base bias

current will flow if a transistor using either the maximum or the minimum h_{FE} values should be used.

The simplest method of establishing the *operating point* of a common-emitter transistor is shown in Fig. 3.12. Applying Kirchhoff's second law to the circuit,

$$E_{cc} = I_b R_1 + V_{BE}$$

where V_{BE} is the base–emitter voltage. Therefore

$$R_1 = \frac{E_{cc} - V_{BE}}{I_b} \tag{3.11}$$

This circuit does not provide any d.c. stabilization against changes in collector current due to change in I_{CBO} or in h_{FE} and its usefulness is limited.

EXAMPLE 3.4

The circuit shown in Fig. 3.13 is designed for operation with transistors having a nominal h_{FE} of 100. Calculate the collector current. If the range of possible h_{FE} values is from 50 to 160, calculate the collector current flowing if a transistor having the maximum h_{FE} is used. Assume $I_{CBO} = 5\ \mu A$ and $V_{BE} = 0.3\ V$.

Solution
From equation (3.11)

$$I_b = \frac{E_{cc} - V_{BE}}{R_1} = \frac{9 - 0.3}{300 \times 10^3} = 29\ \mu A$$

From equation (3.10),

$$I_c = h_{FE} I_b + I_{CEO} = h_{FE} I_b + (1 + h_{FE}) I_{CBO}$$
$$= (100 \times 29) + (101 \times 5)\ \mu A$$

Therefore

$$I_c = 3.405\ mA \qquad (Ans.)$$

Using a transistor of $h_{fe} = 160$,

$$I_c = (160 \times 29) + (161 \times 5)\ \mu A = 5.445\ mA \qquad (Ans.)$$

In the above example the effect of the increased collector current would be to move the operating point along the d.c. load line, and this would lead to signal distortion unless the input signal level were reduced.

A *better bias arrangement*, shown in Fig. 3.14, is to connect a bias resistor R_1 between the collector and base terminals of the transistor. The bias resistor R_1 provides a path for the alternating component of the collector current to feed into the base circuit. This applies *negative feedback* to the circuit and should this not be required the bias circuit is decoupled as shown in Fig. 3.15. The circuit provides some degree of d.c. stabilization against changes in the design value of the collec-

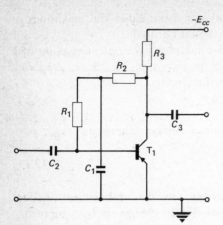

Fig. 3.15 Collector-base bias decoupled to prevent negative feedback

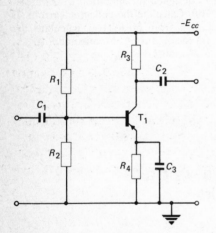

Fig. 3.16 Potential-divider bias

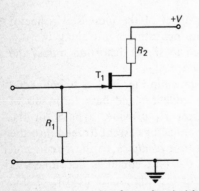

Fig. 3.17 Junction f.e.t. simple bias

tor current, the operation being briefly as follows. An increase in the collector current produces an increased voltage drop across the collector load resistor R_2. This causes the collector-emitter voltage to fall, and since this voltage is effectively applied across the base resistor R_1, the base bias current falls also. The fall in bias current leads to a fall in the collector current which to some extent compensates for the original increase.

For an *improvement in the d.c. stabilization* the bias arrangement of Fig. 3.16 may be employed. The base of the transistor is held at a negative potential V_B by the potential divider $(R_1 + R_2)$ connected across the collector supply, and the emitter is held at a negative potential V_E by the voltage developed across the emitter resistor R_4. The emitter–base bias potential is the difference between V_B and V_E, and the resistor values are chosen so that the junction is forward biased by a fraction of a volt. A base bias current is therefore provided. If negative feedback is not wanted, the emitter resistor R_4 is decoupled by capacitor C_3. D.C. stabilization of the collector current is achieved in the following manner: an increase in the collector bias current, caused by an increase in the temperature of the collector–base junction, is accompanied by an almost equal increase in the emitter current. This results in an increase in the voltage V_E developed across the emitter resistor, and this in turn reduces the forward bias of the emitter–base junction. The base current is reduced causing a decrease in the collector bias current that compensates for the original increase.

F.E.T. Bias

The drain characteristics shown in Fig. 1.12 show that a junction f.e.t. is conducting when the gate–source voltage V_{gs} is zero. The simplest method of biasing an n-channel junction f.e.t. is therefore that given in Fig. 3.17; the disadvantages are (*a*) the maximum input signal amplitude must be very small if excessive distortion is to be avoided, and (*b*) no stabilization against changes in the steady, direct drain current is provided.

Normally the n-channel junction f.e.t. is operated with its gate biased negatively with respect to its source. This can be achieved by the circuit shown in Fig. 3.18. Resistor R_1 connects the gate to the earth line and the voltage drop across R_3 provides the required bias voltage. The gate current is minute and hence, for values of R_1 of a megohm or so, the direct voltage developed across R_1 is negligibly small. Resistor R_3 is often decoupled by means of capacitor C_1 to prevent negative feedback of the signal. This arrangement provides adequate d.c. stability for most small-signal stages provided that the temperature variation is not greater than about 20°C from

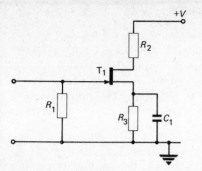

Fig. 3.18 Junction f.e.t. and depletion-type m.o.s.f.e.t. source bias

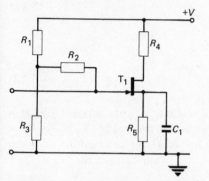

Fig. 3.19 Junction f.e.t. and depletion-type m.o.s.f.e.t. potential-divider bias

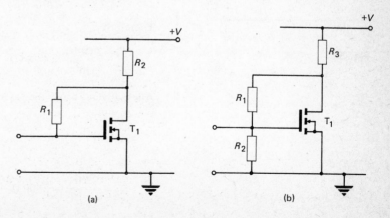

Fig. 3.20 Enhancement-type m.o.s.f.e.t. bias

room temperature. Junction f.e.t.s of the same type are subject to wide spreads in some of their parameters and it may therefore often be necessary to use the more effective bias circuit of Fig. 3.19, which operates in similar fashion to the circuit shown in Fig. 3.16.

If the junction of the bias resistors R_1 and R_3 is directly connected to the gate terminal of the f.e.t. then, at all signal frequencies, the resistors will effectively appear in parallel with the input terminals of the device. The high input impedance of the f.e.t. will then be reduced to a considerably smaller value. To minimize this shunting effect, resistor R_2 can be used to connect the bias resistors to the gate terminal as shown in the figure. The input impedance of the amplifier is then equal to

$$R_2 + R_1 R_3/(R_1 + R_3)$$

and is approximately equal to R_2 since R_2 is chosen to be 1 MΩ or more.

An n-channel depletion-mode m.o.s.f.e.t. must be biased so that its gate is held at a negative potential relative to its source and hence the bias circuits shown in Figs. 3.18 and 3.19 can be employed. An n-channel enhancement mode m.o.s.f.e.t. must be operated with its gate at a positive potential with respect to its source and so a different bias circuit is necessary. The circuit shown in Fig. 3.20a can be used if the operating point $V_{GS} = V_{DS}$ is suitable. If, for reason of obtaining maximum output voltage or minimizing distortion, some other operating point is required then the circuit given in Fig. 3.20b must be used. With this circuit

$$V_{GS} = V_{DS} R_2/(R_1 + R_2)$$

Both circuits provide d.c. stabilization of the operating point in a similar manner to that previously described for transistor collector–base bias.

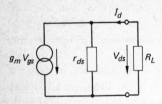

Fig. 3.21 F.e.t. equivalent circuit

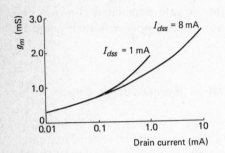

Fig. 3.22 Variation of f.e.t. mutual conductance with drain current

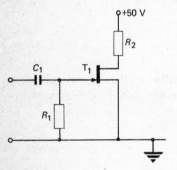

Fig. 3.23

Voltage Gain of F.E.T. Amplifier

At audio-frequencies the performance of a f.e.t. can be described by its mutual conductance $g_m = I_d/V_{gs}$, where I_d and V_{gs} are a.c. components, and the equivalent circuit of Fig. 3.21. The resistance r_{ds} is the output resistance of the device and is equal to $\delta V_{ds}/\delta I_d$ with V_{gs} constant. The output voltage V_{ds} is given by

$$V_{ds} = \frac{g_m V_{gs} R_L r_{ds}}{R_L + r_{ds}}$$

Therefore the voltage gain A_v is given by

$$A_v = \frac{V_{ds}}{V_{gs}} = \frac{g_m R_L r_{ds}}{r_{ds} + R_L} \qquad (3.12)$$

If, as is usual, $r_{ds} > R_L$,

$$A_v = g_m R_L \qquad (3.13)$$

The mutual conductance of a f.e.t. is not constant but is a function of the drain current as shown in Fig. 3.22. As for a transistor, the operating point must be selected to give the required value of g_m. (I_{dss} is the drain current for $V_{gs} = 0$.)

EXAMPLE 3.5

Calculate the drain load resistance required to give the circuit of Fig. 3.23 a voltage gain of 20. The f.e.t. used has $g_m = 4 \times 10^{-3}\,\text{S}$ and $r_{ds} = 100\,\text{k}\Omega$.

Solution
From equation (3.12),

$$A_v = 20 = \frac{4 \times 10^{-3} \times 10^5 R_2}{10^5 + R_2}$$

Therefore

$$R_2 = \frac{20 \times 10^5}{380} = 5.26\,\text{k}\Omega \qquad (Ans.)$$

Alternatively, using the approximate expression given by equation (3.13),

$$A_v = 20 = 4 \times 10^{-3} R_2$$

or

$$R_2 = \frac{20}{4 \times 10^{-3}} = 5000\,\Omega \qquad (Ans.)$$

Voltage, Current and Power Amplifiers

When an amplifier is to be provided between a source and a load or two or more stages are to be connected in cascade, the input and output impedances of the amplifier stages must be chosen to suit the intended application of the amplifier, i.e. a voltage, a current, or a power gain.

(a) VOLTAGE GAIN

A voltage-amplifying stage should have an input impedance that is high compared with the source impedance, and an output impedance which is low compared to the load impedance. The reason for this is best explained by way of a numerical example. Suppose an amplifier having an open-circuit voltage gain of 100 is to be used to amplify a 0.1 V signal from a source of impedance 1000 Ω and deliver the amplified voltage to a 1000 Ω load. If the amplifier has input and output impedances of 50 Ω and 10 000 Ω respectively, then (Fig. 3.24a) the voltage V_{IN} appearing across the input terminals of the amplifier is equal to

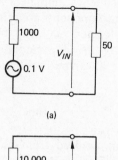

(a)

$$\frac{0.1 \times 50}{1000 + 50} = 4.762 \text{ mV}$$

The amplified open-circuit voltage is 476.2 mV. Thus (Fig. 3.24b)

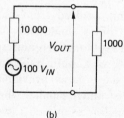

(b)

$$V_{OUT} = \frac{0.4762 \times 1000}{11\,000} = 0.433 \text{ mV}$$

The output voltage developed across the load is smaller than the original signal source voltage and it is evident that an amplifier having an input impedance smaller than its source impedance, and an output impedance larger than its load impedance, is not suited to voltage amplification.

Now consider an amplifier having the same open-circuit voltage gain as before but with its input and output impedance values reversed, i.e. $R_{IN} = 10\,000 \,\Omega$ and $R_{OUT} = 50 \,\Omega$. Now (Figs. 3.24c and d)

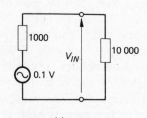

(c)

$$V_{IN} = \frac{0.1 \times 10\,000}{11\,000} = 0.091 \text{ V} \quad \text{and}$$

$$V_{OUT} = \frac{0.091 \times 100 \times 1000}{1050} = 8.658 \text{ V}$$

Thus the use of this amplifier has boosted the signal level 86.58 times. Clearly, the requirement for a voltage-amplifying stage is that its input impedance must be high compared to the impedance of its signal source, and its output impedance must be much smaller than its load impedance.

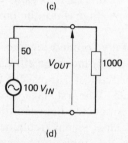

(d)

Fig. 3.24

(b) CURRENT GAIN

An amplifier intended for the amplification of a signal current, rather than a signal voltage, should be designed to have an input impedance which is much smaller than the source impedance and an output impedance that is many times larger than the intended load impedance.

(c) POWER GAIN

For maximum power to be transferred from a source to a load, the load and source impedances should be equal [TS II]. Generally, the impedances of the source and the load will not be equal to the input and output impedances of the amplifier. Maximum power gain will then require the use of transformers to achieve the necessary impedance transformations. Most modern small-signal audio-frequency amplifiers avoid the use of transformers because these components are relatively bulky, heavy and expensive. Transformer matching for maximum power gain is used mainly for the line amplifiers in audio-frequency amplified telephone lines and in the output stage of many a.f. power amplifiers.

Multi-stage Amplifiers

Very often the voltage gain required from an amplifier is greater than that available from a single stage. Then two or more stages must be cascaded to obtain the required voltage gain.

The overall voltage gain A_v of a multi-stage amplifier is the product of the individual stage gains provided the output impedance of each stage is much smaller than the input impedance of the next stage. For example, a two-stage amplifier has an overall gain $A_v = A_1 A_2$ and the gain of an n-stage amplifier is $A_v = A_1 A_2 \ldots A_n$. The signal voltage is applied to the input terminals of the first stage, amplified, and the amplified signal is applied to the input terminals of the next stage and so on for each stage in the amplifier. The output signal of one stage is the input signal of the next and so means of coupling stages together is necessary. The overall current gain is not equal to the product of the current gains of each stage since the output current of one stage is not the input current of the next.

Amplifier stages are normally coupled together by means of a coupling capacitor; when this is inconvenient direct coupling is generally employed. The circuit of a transistor R-C coupled amplifier is shown in Fig. 3.25. The first stage is coupled to the second by the capacitor C_3 connected between the collector of T_1 and the base of T_2. At most signal frequencies the reactance of C_3, and also of the input and output coupling capacitors C_1 and C_5, is negligibly small and can be ignored.

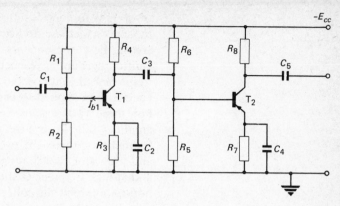

Fig. 3.25 *R-C coupled transistor amplifier*

The overall voltage gain of the amplifier is then at its maximum value.

Although accurate calculation of the overall current and voltage gains, and of the input and output impedances, is rather complicated and lengthy, a reasonably good estimate of these quantities can fairly easily be made.

Consider the circuit of Fig. 3.25 and suppose that T_1 and T_2 are identical transistors having $h_{fe} = 120$ and $h_{ie} = 1000$ ohms. If h_{oe} and h_{re} can both be neglected, the current gain and the input resistance of each transistor is equal to 120 and 1000 ohms respectively. Let $R_4 = R_8 = 2000$ ohms and suppose the bias resistors are of sufficiently high resistance to have negligible shunting effect upon the signal path.

Then, using equation (3.4), the voltage gain of the second stage is $120 \times 2000/1000 = 240$. T_1 has an effective collector load resistance of R_4 in parallel with the input resistance of T_2, i.e. 2000 Ω in parallel with 1000 Ω or 667 Ω. Thus, the voltage gain of the stage is $120 \times 667/1000 = 80$ and so the overall voltage gain is $80 \times 240 = 19\,200$.

All the a.c. component of the collector current of T_1 does not flow into the base of T_2; some of it flows via R_4 to earth instead. The coupling circuit introduces a loss equal to

$$R_4/(R_4 + h_{ie}) \quad \text{or} \quad 2000/(2000 + 1000) = 0.67$$

The overall current gain is therefore

$$120 \times 0.67 \times 120 = 9648$$

The input resistance of the amplifier is the input resistance of the first stage, i.e. 1000 Ω.

The output resistance of the amplifier is equal to the output resistance of T_2 in parallel with R_8. Since h_{oe} is small, the output resistance of the amplifier is large and so the output resistance of the amplifier is approximately equal to R_8 or 2000 Ω.

At low frequencies the reactances of the three coupling capacitors increase and some of the signal voltage is dropped across these components. In addition, the reactances of the emitter decoupling capacitors also increase and some signal frequency current flows in the emitter resistors. Negative feedback is then applied to each stage and their voltage gains fall. For these two reasons the voltage gain of an R-C coupled amplifier falls at low frequencies. The gain also falls at high frequencies; there are two reasons for this: (a) various stray and transistor capacitances are in parallel with the signal path and at high frequencies their reactance becomes low enough to noticeably shunt the collector load impedances; (b) the current gain h_{fe} of the transistor falls with increase in frequency.

For audio-frequency amplifiers, however, the loss of gain due to the transistor itself can be eliminated by the use of a suitable type of transistor in which the decrease in current gain does not occur until frequencies well beyond the audio range are reached. The gain/frequency characteristic of a transistor R-C coupled amplifier is shown in Fig. 3.26. The frequency axis has been plotted to a logarithmic scale so that equal distances on the axis represent equal frequency ratios. This

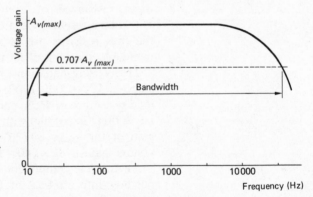

Fig. 3.26 Gain/frequency characteristic of an R-C coupled transistor amplifier

practice gives a better idea of the usefulness of the amplifier for the amplification of sound waveforms since the human ear is sensitive to frequency ratios rather than absolute frequency changes.

The bandwidth of an amplifier is usually quoted as the frequency band between the 3 dB points; these are the two frequencies, one high and one low, at which the voltage gain has fallen by 3 dB (or to 0.707 times) from the gain at middle frequencies. A reasonably accurate estimate of the bandwidth of a multi-stage amplifier is not easily obtained since it depends upon a number of interacting time constants. The lower

3 dB frequency is usually only a few tens of hertz while the upper 3 dB frequency is some tens or hundreds of kilohertz. Little error is therefore introduced by taking the upper 3 dB frequency as the bandwidth of the amplifier. This can be quickly measured using equation (3.18).

Fig. 3.27 shows an *R-C* coupled f.e.t. amplifier; clearly the circuit is very similar to that of a transistor *R-C* coupled amplifier, as is its gain/frequency characteristic.

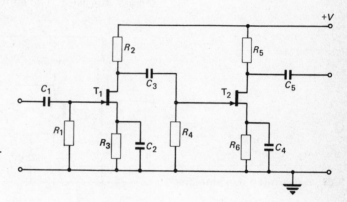

Fig. 3.27 F.e.t. *R-C* coupled amplifier

The relative advantages of transistors and f.e.t.s have been mentioned in Chapter 1. The f.e.t. has a very much higher input impedance than the transistor but the transistor has the greater mutual conductance and hence its voltage gain is higher for a given value of load resistance. In many cases, the best results can be obtained with the hybrid circuit shown in Fig. 3.28. The gain of this circuit will fall at both high and low frequencies for the same reasons as previously discussed for transistor amplifiers.

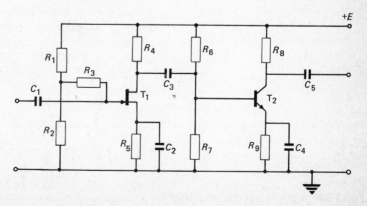

Fig. 3.28 A hybrid *R-C* coupled amplifier

Direct Coupling

The high reactance of the coupling capacitor at very low frequencies prevents an R-C coupled amplifier from handling very low frequency and direct signals. The low-frequency range of an amplifier can only be extended to zero frequency if the use of reactive coupling components is eliminated. Amplifiers in which stages are directly coupled are known as direct-coupled or d.c. amplifiers.

If two transistors are directly coupled together the collector of the first will be at the same potential as the base of the second transistor. The emitter potential must then be adjusted,

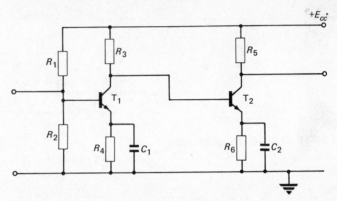

Fig. 3.29 n-p-n/n-p-n d.c. amplifier

by suitable choice of the emitter resistance and the emitter current, to be slightly less than the base potential in order to obtain the correct base/emitter voltage. If, as in Fig. 3.29, n-p-n (or p-n-p) transistors are directly coupled together the collector voltages of successive stages will become larger and larger. The d.c. voltage level at the output terminals will therefore be greater than the input level; for some applications this increase in the d.c. voltage level will not matter, but for others it is necessary that the input and output levels are approximately equal. Also, of course, moving the operating point upwards along the load line will restrict the possible output voltage variation. The increase in d.c. voltage level can be avoided if a combination of n-p-n and p-n-p transistors is used as shown in the circuit of Fig. 3.30.

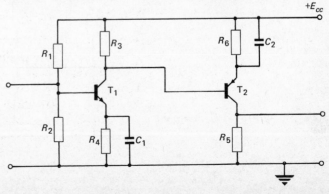

Fig. 3.30 n-p-n/p-n-p d.c. amplifier

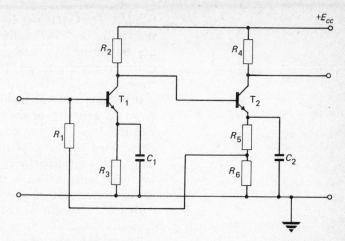

Fig. 3.31 A d.c. feedback pair amplifier

A commonly employed circuit, in which the bias voltage for the first stage is derived from the second stage is shown in Fig. 3.31. The base of transistor T_2 is at the same potential as the collector of T_1 while the base of T_1 is held at the potential at the junction of the T_2 emitter resistors R_5 and R_6. This arrangement has two advantages over the circuits given in Figs. 3.29 and 3.30; the input impedance is not reduced by the shunting effect of the bias resistors R_1 and R_2, and improved d.c. stabilization is provided.

The major difficulty with d.c. amplifiers is the reduction of drift. Any small fluctuations, or drifts, of voltage or current in the first stage in an R-C coupled amplifier are not passed on to the second stage by the coupling network. In a d.c. amplifier, however, such drifts act as spurious signals and are amplified by the following stages. A transistor amplifier, for example, is unable to distinguish between a signal of, say, 2 mV and a 2 mV change in the emitter/base bias voltage which could occur because of a change in temperature.

The causes of drift in a d.c. amplifier are changes in component values and f.e.t. or transistor characteristics with voltage and/or with time, and changes in the power supplies. Many circuits have been devised to reduce drift to a minimum.

Wideband Amplifiers

Wideband amplifiers are used in telecommunication and electronic systems to amplify both analogue and digital signals. The wideband analogue signal might be a television picture signal occupying a bandwidth of 0–5.5 MHz, or perhaps a multi-channel telephony signal which occupies a bandwidth that depends upon the number of channels involved. For example, the most modern system occupies the frequency band of 4–60 MHz, while a basic 12-channel group occupies the band 60–108 kHz. Increasingly nowadays, digital signals are transmitted by telecommunication systems and may require amplification, e.g. pulse modulation systems and data transmission [TS II].

If the gain of a wideband amplifier is required to be constant down to zero frequency, then the use of reactive coupling elements must be avoided. The high frequency response is determined by the f_t of the transistors and the various stray and transistor capacitances in the circuit. The choice of transistor type must be such that the transistor current gain is constant over the required bandwidth, and then the upper 3 dB frequency will be determined only by the circuit capacitances.

The voltage gain of a stage will have fallen by 3 dB from its maximum medium-frequency value at the frequency at which the reactance of the total shunt capacitance is equal to the effective load resistance. Thus, if the effective load resistance is reduced in value by using a lower collector load resistor, the upper 3 dB frequency will be moved to a higher value. Unfortunately, reducing the collector load resistance will also reduce the voltage gain of a stage, and perhaps necessitate the use of an extra stage in order to obtain a required overall voltage gain. This, however, can lead to further problems since the addition of an extra stage not only increases the overall gain but also reduces the overall 3 dB bandwidth.

It is possible to increase the bandwidth of an amplifier stage without reducing the collector load resistance. One method, which will be dealt with in the next chapter, is to apply *negative feedback* to the circuit. Another method, known as *shunt compensation*, is to connect an inductor of suitable value in series with the collector or drain resistance. Fig. 3.32 shows a f.e.t. shunt compensated circuit. Coupling capacitors have been shown but would have been omitted if the circuit were expected to respond to very low frequency signals. At most signal frequencies the reactance of the inductor is negligibly small compared with the collector load resistance and its inclusion in the circuit has no effect on the gain of the circuit. At high frequencies the reactance of the inductor is no longer negligible and the impedance of the drain load increases with increase in frequency. The effect on the gain/frequency response of the amplifier is illustrated by Fig. 3.33. When the correct amount of inductance has been added the condition known as *maximal flatness* is obtained in which the gain is flat over the widest possible bandwidth. If a larger value of inductance is used, the gain of the amplifier will rise to a maximum at some high frequency.

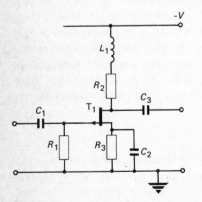

Fig. 3.32 Shunt compensated f.e.t. amplifier

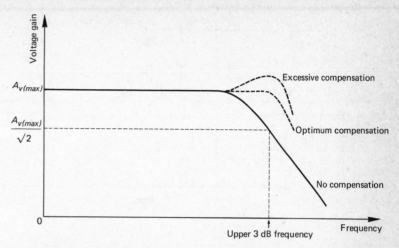

Fig. 3.33 Gain/frequency characteristic of a shunt-compensated amplifier

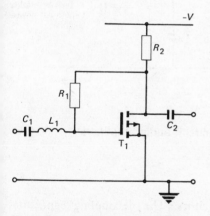

Fig. 3.34 Series-compensated m.o.s.f.e.t. amplifier

The gain at high frequencies can also be improved by the use of *series compensation* in which an inductor is connected in series with either the input or the output of the stage (see Fig. 3.34).

Integrated Circuit Amplifiers

Audio-frequency amplifiers are available in integrated circuit form that can be used as a replacement for the discrete component circuits described earlier in this chapter. Some devices are manufactured purely as a.f. amplifiers whilst others, known as *operational amplifiers*, can be used for a number of different purposes. Since operational amplifier circuits require the application of negative feedback they will be described in the next chapter. An integrated audio amplifier is designed to minimize the need for coupling and decoupling capacitors but their use cannot be completely eliminated. Since only small values of capacitance can be provided within the silicon chip, this means that the external capacitors must be fitted to the appropriate terminals of an i.c. amplifier. It is also inconvenient to provide large values of resistance within an i.c. and these will also be provided externally. Such resistors will normally be acting as bias or load resistors.

Some typical examples of integrated audio amplifiers are shown in Fig. 3.35. Fig. 3.35a shows an i.c. amplifier having a number of external resistors and capacitors fitted to its terminals. Capacitors C_2 and C_3 decouple bias resistors R_2 and R_3; C_5 decouples the power supply; and C_1 and C_4 are the input

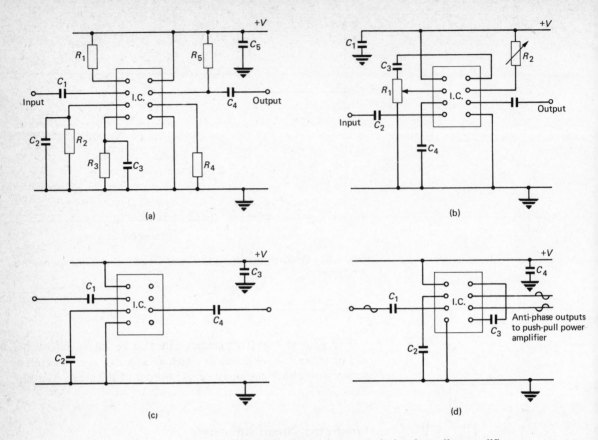

Fig. 3.35 Some integrated circuit audio amplifiers

and output coupling capacitors. The decoupling capacitors have a value of about 25 μF and the coupling capacitors are about 1 μF; these values are much larger than can be provided within the integrated circuit. R_1 and R_4 are also bias resistors but do not need to be decoupled and R_5 is a load resistor. A different i.c. amplifier circuit is given in Fig. 3.35b. The gain of the amplifier can be varied by adjustment of resistor R_2. The gain/frequency characteristic of the amplifier is determined by the series circuit R_1-C_3 and can be adjusted by means of variable resistor R_1. Capacitor C_4 is the only decoupling capacitor required by this integrated circuit and C_1 decouples the power supply. Some integrated circuits have been designed to minimize the number of external components necessary and Fig. 3.35c shows one that needs only one decoupling capacitor and no resistors in addition to the input and output coupling capacitors. Finally, Fig. 3.35d shows a circuit designed to act as the driver stage for a push–pull power amplifier.

The important characteristics of an integrated audio-frequency amplifier are its gain, its maximum peak-to-peak output voltage, its 3 dB bandwidth, and its input and output impedances. One i.c. which is in common use has a gain of 40 dB, a maximum output voltage swing of 8 V when a 10 V power supply is used, and a bandwidth of 75 kHz. The input resistance is 100 kΩ and the output resistance is 150 Ω. Another device has 46 dB gain and a maximum peak-peak output voltage of 2.5 V from a 6 V power supply. The input impedance is 1000 Ω and the output impedance is 1.5 Ω.

An integrated circuit wideband amplifier will have a flat frequency characteristic over a bandwidth of, typically, 0–10 MHz with an upper 3 dB frequency of about 50 MHz.

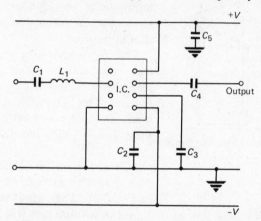

Fig. 3.36 Integrated wideband amp-lifier

Negative feedback can be applied to further increase the upper 3 dB frequency, or series compensation can be used. Fig. 3.36 shows the circuit of an integrated wideband amplifier with series compensation. A circuit of this type might have a 3 dB bandwidth of 80 MHz and a gain of 30 dB.

Measurements on Audio-frequency Amplifiers

A number of amplifier parameters can be measured using a signal generator and a c.r.o. The more important of these are

> (a) the detection of distortion
> (b) the measurement of gain
> (c) the determination of the input and output impedances
> (d) square wave handling ability.

(a) THE DETECTION OF DISTORTION
The final stage in an amplifier handles the largest amplitude signal and it is therefore in this stage that signal distortion is most likely to occur. If the amplitude of the signal voltage

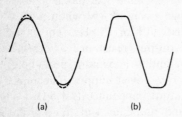

(a)   (b)

Fig. 3.37 Waveform distortion

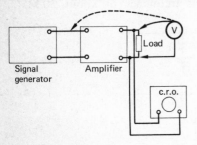

Fig. 3.38 Circuit for measurement of amplifier gain

applied to the output transistor is excessive, the output current and voltage swings will be large and will encroach into the non-linear regions of the output characteristics. The effect of this is to make the output signal waveform differ from the input waveform. With a sinusoidal signal this distortion becomes noticeable as a flattening of either or both of the positive and negative peaks (see Fig. 3.37a).

If the amplitude of the input signal is still further increased, the output transistor may be driven into cut-off during one half-cycle and into saturation in the other half-cycle. The signal waveform is then clipped and has the appearance shown in Fig. 3.37b.

To determine the voltage which, when applied to the input terminals of an amplifier, just causes distortion of the output waveform to become apparent, the test circuit of Fig. 3.38 should be used.

The frequency of the generator should be set to the desired test value and then the generator output voltage should be steadily increased from zero until the onset of distortion is noticed. The maximum input signal amplitude that the amplifier can handle without distortion is then somewhat less than the value at which the distortion becomes evident.

(b) THE MEASUREMENT OF AMPLIFIER GAIN

The gain of an amplifier can be measured using the test circuit of Fig. 3.38. The output voltage of the signal generator should be set to a value that will not produce distortion of the output signal waveform, and the generator frequency should be adjusted to the required test frequency. The gain of the amplifier is then the ratio output-voltage/input-voltage; these two voltages should preferably be measured with the same voltmeter since then the measurement will not be affected by meter accuracy.

Often the gain/frequency characteristic of an amplifier is to be measured. The procedure is then to keep the voltage applied to the amplifier constant while the generator frequency is varied, in a number of steps, over the bandwidth. At each step the output voltage should be measured and the gain calculated. The results are then plotted on suitable graph paper to obtain the gain/frequency characteristic of the amplifier.

(c) DETERMINATION OF INPUT AND OUTPUT IMPEDANCES

To measure the input impedance of an amplifier, the input voltage delivered by the signal generator to the amplifier is set to some convenient value (less than the value which causes

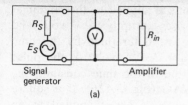

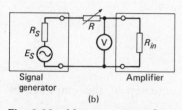

Fig. 3.39 Measurement of the input impedance of an amplifier

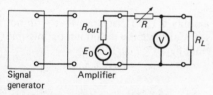

Fig. 3.40 Measurement of the output impedance of an amplifier

distortion of the output waveform). Then, referring to Fig. 3.39*a*

$$V = \frac{E_s R_{IN}}{R_s + R_{IN}} \qquad (3.14)$$

Here, R_s is the output impedance, and E_s is the e.m.f. of the signal generator, while R_{IN} is the amplifier input resistance to be determined.

A variable resistance R is then connected in series with the input terminals of the amplifier (Fig. 3.39*b*), and its value is increased until the input voltage has fallen to one half its original value. Then,

$$\frac{V}{2} = \frac{E_s R_{IN}}{R_s + R + R_{IN}} \qquad (3.15)$$

From equations (3.14) and (3.15)

$$(R_s + R_{IN})2 = R_s + R + R_{IN}$$

Therefore

$$R_{IN} = R - R_s \qquad (3.16)$$

If the input impedance of the amplifier is not purely resistive at the frequency of measurement, the magnitude of the impedance is obtained.

The output impedance of an amplifier can be measured using a similar technique, the arrangement being shown in Fig. 3.40. Then, $R_{OUT} = R - R_L$. Care must be taken to ensure that the addition of resistance in series with the load does not upset the operating conditions of the amplifier and alter the output impedance of the amplifier.

(*d*) SQUARE WAVE HANDLING ABILITY

An alternative method of testing the performance of an audio-frequency amplifier is to apply a square wave to its input terminals and observe the output waveform on a c.r.o. If the waveform generator is set to a lower audio frequency, such as 400 Hz, a reasonable square wave output should be observed if the amplifier has a bandwidth extending from about 25 Hz up to about 20 kHz. If the bandwidth of the amplifier is limited at either low or high frequencies some waveform distortion will be present.

The fall-off in the gain of an *R-C* coupled amplifier at low and at high frequencies is produced, by circuits of the form shown in Figs. 3.41*a* and *b*. In Fig. 3.41*a*, capacitor *C* represents the coupling capacitor, while in Fig. 3.41*b* *C* represents the total stray and transistor capacitance which shunts the signal path.

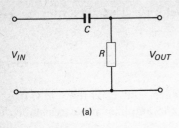

(a)

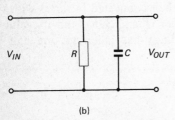

(b)

Fig. 3.41 Circuits determining (a) the lower and (b) the upper 3 dB frequency of an amplifier stage

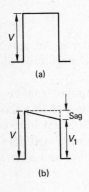

(a)

(b)

(c)

Fig. 3.42 Showing (a) an ideal pulse, (b) a pulse with sag, and (c) a pulse with risetime t

When a voltage step of V volts is applied to the circuit of Fig. 3.41a the output voltage rises abruptly to V volts and then decays exponentially at a rate determined by the time constant CR seconds, i.e. $V_{OUT} = Ve^{-t/RC}$ volts. If the pulse duration is an appreciable fraction of the time constant, the output waveform will exhibit SAG (Fig. 3.42b). If a square wave of pulse repetition frequency f Hz is applied to an amplifier, the percentage sag measured

$$\frac{V - V_1}{V} \times 100\%$$

is related to the lower 3 dB frequency of the amplifier by equation (3.17), i.e.

$$\% \text{ sag} = \frac{\pi f_{3dB}}{f} \tag{3.17}$$

This means that the low-frequency performance of an amplifier can be rapidly assessed by means of a square wave test.

EXAMPLE 3.6

An amplifier is to be used to amplify a 100 Hz square wave with not more than (i) 5% sag, (ii) 1% sag. Calculate the maximum permissible lower 3 dB frequency of the amplifier.

Solution

(i) From equation (3.17) $f_{3dB} = \dfrac{5 \times 100}{\pi} = 159 \text{ Hz}$ (*Ans.*)

(ii) $f_{3dB} = \dfrac{1 \times 100}{\pi} = 32 \text{ Hz}$ (*Ans.*)

When a step voltage of V volts is applied to the circuit given in Fig. 3.41b the output voltage cannot immediately rise to V volts but rises exponentially with a time constant CR seconds, i.e.

$$V_{OUT} = V(1 - e^{-t/CR}) \text{ volts}$$

The RISETIME of the output waveform is the time taken by the output voltage to rise from 10% to 90% of its final value of V volts (see Fig. 3.42c). The risetime of an amplifier is related to its upper 3 dB frequency according to equation (3.18), i.e.

$$\text{Risetime} = \frac{0.35}{f_{3dB}} \tag{3.18}$$

EXAMPLE 3.7

The risetime of an amplifier is measured as 200 nS. Determine its upper 3 dB frequency.

Solution

From equation (3.18), $f_{3dB} = \dfrac{0.35}{200 \times 10^{-9}} = 1.75\ \text{MHz}$ (*Ans.*)

Exercises

3.1. Draw circuit diagrams showing how a junction transistor may be used in a single-stage amplifier (*a*) with common base, (*b*) with common emitter.

Derive an approximate relationship, for small input signals, between the current gain in the common-base and common emitter circuits for the two methods of connection.

A transistor connected in common emitter shows changes in collector and emitter currents of 1.0 mA and 0.98 mA respectively. What change in base current produces these, and what is the current gain of the transistor? (C&G)

3.2. For a transistor used in the common-emitter configuration the relationships between collector current and collector voltage, with fixed values of base current, are given in Table A.

Table A

Collector–emitter voltage (V)	Collector current (mA)				
	I_b $= -30\ \mu\text{A}$	$-50\ \mu\text{A}$	$-70\ \mu\text{A}$	$-90\ \mu\text{A}$	$-110\ \mu\text{A}$
−2	−0.9	−1.55	−2.2	−2.85	−3.55
−4	−0.92	−1.65	−2.4	−3.05	−3.77
−6	−0.95	−1.77	−2.55	−3.25	−4.0
−8	−0.98	−1.9	−2.75	−3.5	−4.2

Draw the static characteristics of the transistor and use these to determine the current gain when the collector voltage is −5 V.

The transistor is to be used as a common-emitter amplifier with a load resistance of 1800 Ω and a collector battery voltage of −9 V. Draw the load line and use this to find the base current for a collector voltage of −4 V. (C&G)

3.3. (*a*) Sketch the circuit diagram of a single-stage Class A amplifier in which the active device is a f.e.t. (*b*) Describe how the operating point will alter with variation in the temperature of the device. (*c*) Explain the principle of operation of the particular type of transistor used in your diagram. (C&G)

3.4. Draw the circuit diagram of an audio amplifier in which the active device is an integrated circuit. List the function of each external component drawn and say why it could not be provided within the integrated circuit.

3.5. An audio-frequency amplifier is to be constructed using either a junction field effect transistor having a mutual conductance of 2 mS or a bipolar transistor having an input resistance of

$1500\,\Omega$. The amplifier is to provide a drain or collector a.c. current of $1\,mA$ peak value in a load resistance of $3\,k\Omega$. Determine (i) the voltage gain of each amplifier, (ii) the signal e.m.f. required to produce the desired output voltage if the resistance of the source is $30\,k\Omega$. Comment on your answer.

3.6. Draw the circuit diagram of a two-stage $R\text{-}C$ coupled audio-frequency amplifier using transistors. Show typical values for the components and suitable bias and stabilization arrangements. What factors influence the frequency response of the amplifier you have described? (C&G)

3.7. Describe, briefly, with reference to circuit diagrams, two methods of interstage coupling used in audio-frequency amplifiers. Sketch typical gain/frequency characteristics for each type and explain why a logarithmic frequency scale is normally used for this purpose. (C&G)

3.8. Explain the operation of a resistance loaded transistor amplifier in the common-emitter configuration. Draw the circuit diagram of a two-stage $R\text{-}C$ coupled audio-frequency amplifier in the common-emitter configuration. Show clearly (a) the biasing and stabilization arrangements, (b) the directions of the currents flowing, (c) suitable component values.

3.9. Explain, with the aid of circuit diagrams, how you would measure (i) the sensitivity and (ii) the gain/frequency characteristic of an amplifier. In a measurement of the gain/frequency characteristic the data given by Table B was obtained with the input voltage maintained constant at $20\,mV$.

Table B

Frequency (kHz)	0.03	0.1	0.3	1	3	10	30	100
Voltage output (V)	1.6	2.5	3.0	3.1	3.1	3.0	2.8	1.5

Plot the gain/frequency characteristic of the amplifier and determine its $3\,dB$ bandwidth. How could the upper $3\,dB$ frequency be extended to a higher figure?

3.10. Draw the following sets of characteristics, labelling the axes with typical figures: (i) the drain characteristics of a p-channel m.o.s.f.e.t. (say which type you have drawn), (ii) the output characteristics of an n-p-n transistor.

For one of these characteristics, explain how you would draw the d.c. and a.c. load lines corresponding to particular load resistances. Show how the a.c. load line can be employed to determine the voltage or the current gain of the device.

3.11. A common-emitter connected transistor has the data given in Table C.

Plot the output characteristics of the transistor. The transistor is used in a single-stage amplifier having an emitter resistance of $1000\,\Omega$ and a collector load resistance of $2000\,\Omega$. The collector supply voltage is $18\,V$. The output terminals of the amplifier are connected to a load resistance of $2000\,\Omega$.

(a) Draw the d.c. load line and select a suitable operating point. (b) Draw the a.c. load line. (c) Calculate the current gain of the circuit when a $30\,\mu A$ peak signal is applied to the base.

3.12. Explain the reasons why a transistor or a f.e.t. amplifier is usually provided with some degree of d.c. stabilization. Draw the circuit diagram of the potential-divider bias circuit when used with (a) a n-p-n bipolar transistor and (b) an n-channel

Table C

	$I_c(mA)$		
V_{ce} (V)	$I_b = 20\,\mu A$	$50\,\mu A$	$80\,\mu A$
2	1.7	3.25	4.9
4	1.95	3.6	5.3
6	2.1	3.8	5.5
8	2.25	4.0	5.7
10	2.4	4.2	5.9

enhancement-type m.o.s.f.e.t. For (b) explain how the bias circuit operates.

3.13. Draw the circuit diagram of a two-stage R-C coupled f.e.t. circuit. State the types of f.e.t. you have drawn and explain why the gain of the circuit falls at both low and high frequencies.

3.14. (a) Explain the differences between the terms "pinch-off region" and "pinch-off voltage" in a field-effect transistor. (b) Why does the bias arrangement used in the Class A amplifier shown in Fig. 3.19 give good d.c. stability? What is the purpose of R_2? The f.e.t. used has a mutual conductance of 3 mS and the circuit output is connected to an external load of 3 kΩ. If $R_1 = 47$ kΩ, $R_2 = 1$ MΩ, $R_3 = 5.1$ kΩ, $R_4 = 1$ kΩ, $R_5 = 1$ kΩ determine (i) the voltage gain, (ii) the circuit a.c. input resistance.

3.15. Draw the circuit diagram of a junction f.e.t. amplifier stage which uses potential-divider bias. State the function of each component drawn and describe how the bias arrangement operates to provide d.c. stability.

Short Exercises

3.16. Describe, with the aid of a suitable mutual characteristic, what is meant by Class A operation of a transistor or a f.e.t.

3.17. Draw the circuit diagram of a two-stage resistance-capacitance coupled common-emitter transistor amplifier in which the first stage uses a p-n-p transistor and the second stage uses an n-p-n transistor.

3.18. A 3-stage transistor amplifier has a voltage gain of 1000. The maximum output signal voltage is 9 V. Calculate (i) the sensitivity of the amplifier, (ii) the gain of each of the three identical stages.

3.19. A junction transistor has an input resistance of 2000 Ω and a mutual conductance of 50 mS. What are (i) its short-circuit current gain, (ii) its peak-to-peak collector current?

3.20. What are the differences between a voltage amplifier, a current amplifier, and a power amplifier?

3.21. Draw the circuit of a single-stage amplifier using an n-channel depletion-type m.o.s.f.e.t. Explain how bias is applied to the f.e.t. and how d.c. stability is provided.

3.22. Draw the circuit diagram of a two-stage amplifier in which the first stage uses an n-channel enhancement-type m.o.s.f.e.t. and the second stage uses a p-n-p bipolar transistor.

3.23. A junction f.e.t. has $g_m = 4$ mS and $r_{ds} = 125$ kΩ and is connected in a circuit with a drain load resistor of 4.7 kΩ. Calculate the voltage gain of the circuit.

3.24. Make a list of some applications of (i) R-C coupling and (ii) d.c. coupling of amplifier stages.

3.25. What is meant by each of the following terms used in conjunction with an amplifier: (i) sensitivity, (ii) bandwidth, (iii) input impedance?

3.26. Fig. 3.19 shows how a resistor R_2 can be connected to increase the input impedance of a f.e.t. amplifier. Explain why the same technique cannot be used to increase the input impedance of a bipolar transistor stage.

3.27. What would be the input impedances of (i) the ideal voltage amplifier and (ii) the ideal current amplifier?

3.28. List the function of each component shown in the circuit of Fig. 3.35d.

3.29. An amplifier is required to pass a 120 Hz square wave with no more than 8% sag. Determine the lower 3 dB frequency of the amplifier.

4 Negative Feedback

Introduction

An audio-frequency amplifier can be designed to have a certain current, voltage or power gain together with particular values of input and output impedance. The amplifier will add noise and distortion to the signals handled by it. The components employed in the amplifier, both passive (resistors, capacitors, etc.) and active (transistors and f.e.t.s), will vary in value with both time and change in temperature and will have manufacturing tolerances. The gain of the amplifier may therefore vary with time, with change in ambient temperature, and when a component has to be replaced by another of the same type and (nominal) value. The parameters of a transistor, such as its current gain, depend on the d.c. operating conditions and therefore any fluctuations in the power supply may also cause the gain of the amplifier to alter. For many applications a more or less constant gain is necessary, and this can be obtained if *negative feedback* (n.f.b.) is applied to the amplifier, at the expense, however, of a reduction in gain.

Types of Negative Feedback

An n.f.b. amplifier has a fraction of its output signal fed back into its input terminals in *antiphase* with the input signal. Differences exist in the methods used to derive the fed-back signal and to introduce it into the input circuit; these differences lead to the classification of n.f.b. amplifiers into four types: voltage-voltage feedback, voltage-current feedback, current-current feedback, and current-voltage feedback.

Voltage-Voltage Feedback

With voltage-voltage feedback a voltage, that is proportional to the output voltage of the amplifier, is fed back and applied

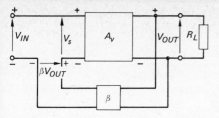

Fig. 4.1 Voltage-voltage negative feedback

in series with the input signal voltage (Fig. 4.1). The amplifier has a gain before feedback of $A_v\underline{/0°}$ and β is the fraction of the output voltage fed back into the input circuit. If the amplifier has an odd number of stages, so that the gain is $A_v\underline{/180°}$, the connections of the β network must be reversed.

The output voltage of the amplifier is V_{OUT}, and so the voltage fed back into the input circuit is βV_{OUT}. The fed-back voltage is arranged to be in antiphase with the input signal V_{IN}. Therefore

$$V_S = V_{IN} - \beta V_{OUT}$$
$$V_{OUT} = A_v V_S = A_v(V_{IN} - \beta V_{OUT})$$
$$= A_v V_{IN} - \beta A_v V_{OUT}$$
$$V_{OUT}(1 + \beta A_v) = A_v V_{IN}$$
$$V_{OUT} = \frac{A_v V_{IN}}{1 + \beta A_v}$$

Therefore

$$\text{Voltage gain with n.f.b.} = A_{v(F)} = \frac{V_{OUT}}{V_{IN}} = \frac{A_v}{1 + \beta A_v} \qquad (4.1)$$

EXAMPLE 4.1

A voltage amplifier has a voltage gain of 100 before n.f.b. is applied. Calculate its voltage gain if 3/100 of the output voltage is fed back to the input in antiphase with the input signal.

Solution
From equation (4.1),

$$A_{v(F)} = \frac{100}{1 + (3/100 \times 100)} = \frac{100}{4} = 25 \quad (Ans.)$$

Thus the application of n.f.b. to an amplifier results in a considerable reduction in gain. An example of voltage-voltage n.f.b. is shown in Fig. 4.2. The common-collector circuit of Fig. 4.2 is commonly known as an *emitter follower*. If the input signal voltage goes positive the base current is reduced and the current gain of the transistor gives an amplified emitter current. This current passes through the emitter resistor and develops a positive-going output voltage, i.e. the output voltage *follows* the input voltage. The output voltage is in series with the input signal voltage and is of opposite polarity; thus *all* the output voltage is applied as n.f.b., i.e. β is unity.

The voltage gain of a transistor amplifier without n.f.b. is

$$A_v = \frac{A_i R_L}{R_{IN}}$$

The effective load resistance $R_{L(eff)}$ seen by the transistor in an emitter follower circuit is the resultant of the emitter

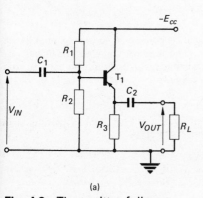

(a)

Fig. 4.2 The emitter follower

resistor R_3 in parallel with the external load resistance R_L. The voltage gain of an emitter follower is

$$A_{v(F)} = \frac{A_i R_{L(eff)}/R_{IN}}{1 + \frac{A_i R_{L(eff)}}{R_{IN}}} = \frac{A_i R_{L(eff)}}{R_{IN} + A_i R_{L(eff)}} \quad (4.2)$$

Clearly the voltage gain is always less than unity.

EXAMPLE 4.2

An emitter follower employs a transistor having a current gain of 100 and an input resistance of 1500 Ω. If the emitter resistance is 2000 Ω calculate the voltage gain of the circuit.

Solution
From equation (4.2)

$$A_{v(F)} = \frac{100 \times 2000}{1500 + 100 \times 2000} = 0.993 \quad (Ans.)$$

The maximum value of emitter resistance is determined by the collector supply voltage and the required maximum output signal voltage. The emitter resistance can be further increased if a second power supply voltage is available, as shown in Fig. 4.3.

The emitter follower has a fairly high input impedance, equal to $h_{fe} R_{L(eff)}$ in parallel with the bias resistor(s), and a low output impedance, of R_s/h_{fe} in parallel with R_3. The circuit is commonly employed to connect a high impedance source to a low impedance load with very little loss of signal voltage.

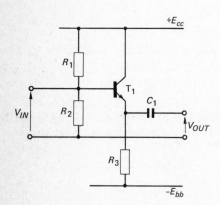

Fig. 4.3 Emitter follower with dual power supply voltages

EXAMPLE 4.3

An emitter follower is to be used to connect a source of impedance 100 kΩ to a 75 Ω coaxial cable. The transistor has a current gain h_{fe} of 150. Calculate the emitter resistance which should be used so that the follower is matched to the cable.

Solution
The output resistance of the circuit is $100 \times 10^3/150$ in parallel with R_3 and is required to be equal to 75 Ω. Therefore

$$75 = \frac{R_3 \times (100 \times 10^3)/150}{R_3 + (100 \times 10^3)/150}$$

$$50 \times 10^3 + 75 R_3 = 666.67 R_3$$

$$R_3 = 84.5 \ \Omega \quad (Ans.)$$

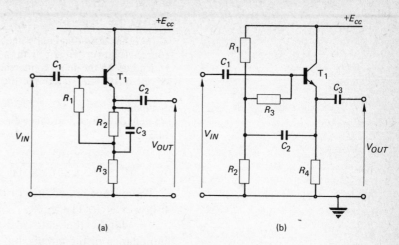

Fig. 4.4 The bootstrapped emitter follower

The high input impedance of the common-collector transistor is shunted by the bias resistors R_1 and R_2. When a very high input impedance is needed the emitter follower must be *bootstrapped* (see Fig. 4.4). In Fig. 4.4*a* the emitter resistance has been divided into two parts R_2 and R_3, whose values are chosen so that the required bias voltage appears at their junction. The junction of the two resistors is connected to the base of T_1 by resistor R_1. Since the voltage gain of an emitter follower is approximately unity, the signal voltages appearing at the two ends of R_1 are very nearly equal. Very little signal-frequency current then flows in R_1 and so its shunting effect on the signal path is extremely small. An alternative arrangement is illustrated by Fig. 4.4*b*. The signal-frequency voltage across the emitter resistor R_4 is approximately equal to the signal voltage at the base of T_1. As a result the a.c. potentials either side of R_3 are very nearly the same and this means that R_3 has a very high a.c. resistance. The d.c. resistance of R_3 must be fairly small because the component must pass the required base bias current. The function of capacitor C_2 is that of a d.c. block and its value should be such that it has negligible reactance at all signal frequencies.

The circuit of a source follower is shown in Fig. 4.5*a*. The voltage gain of a f.e.t. amplifier, before n.f.b. has been applied, is $A_v = g_m R_L$ and hence the voltage gain of a source follower is

$$A_{v(F)} = \frac{g_m R_{L(eff)}}{1 + g_m R_{L(eff)}} \tag{4.3}$$

where $R_{L(eff)}$ is the resistance of the source resistor in parallel with the external load resistance.

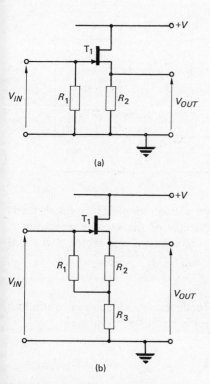

Fig. 4.5 The source follower

EXAMPLE 4.4

A f.e.t. has a mutual conductance g_m of 3.5 mS and is connected in a source follower circuit with a source resistor of 1000 Ω. If the resistance of the load connected across the output terminals of the circuit is 1000 Ω, determine the voltage gain of the circuit.

Solution

The effective load resistance $R_{L(eff)}$ is 500 Ω and hence, from equation (4.3),

$$A_{v(F)} = \frac{3.5 \times 10^{-3} \times 500}{1 + 3.5 \times 10^{-3} \times 500} = 0.636 \quad (Ans.)$$

The output impedance of the f.e.t. is $1/g_m$, and of the source follower is $R_2/(1 + g_m R_2)$.

The larger the gate resistance R_1, the greater the input impedance of the circuit, but on the other hand, increase in R_1 makes the d.c. operating conditions less stable. An alternative arrangement which overcomes this difficulty is shown in Fig. 4.5*b*. The source resistor R_2 is split into two and the gate resistor is connected to their junction.

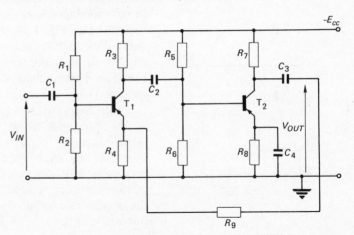

Fig. 4.6 A two-stage voltage-voltage n.f.b. amplifier

Fig. 4.6 shows a two-stage voltage amplifier in which feedback is applied from the collector of the second transistor to the emitter of the first. The voltage feedback factor β is equal to

$$R_4/(R_4 + R_9)$$

Some local *voltage-current* feedback is also applied to the first stage because of its undecoupled emitter resistor.

Voltage-Current Feedback

When voltage-current feedback is applied to an amplifier the fed-back voltage is proportional to the current flowing in the load. This type of feedback is often called current feedback

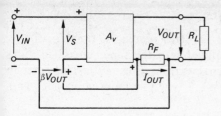

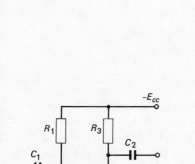

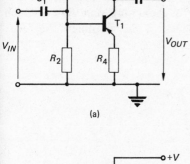

(a)

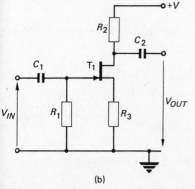

(b)

Fig. 4.7 Voltage-current negative-feedback

Fig. 4.8 Two voltage-current negative-feedback circuits

even though a current is *not* fed back. Fig. 4.7 represents a voltage-current n.f.b. amplifier in which the output current I_{OUT} flows in both the load R_L and the feedback resistor R_F. This current develops a voltage, $I_{OUT}R_F$, that is applied in series with the input signal voltage. The voltage $I_{OUT}R_F$ is the fraction βV_{OUT} of the output voltage fed back, and therefore $\beta V_{OUT} = I_{OUT}R_F$. The output voltage V_{OUT} is given by $I_{OUT}R_L$. Therefore

$$\frac{\beta V_{OUT}}{V_{OUT}} = \frac{I_{OUT}R_F}{I_{OUT}R_L}$$

or

$$\beta = \frac{R_F}{R_L} \qquad (4.4)$$

Two examples of voltage-current feedback are given in Fig. 4.8. It will be seen that the feedback voltage is obtained by leaving the emitter or source resistor undecoupled. The feedback factors of the two circuits are respectively R_4/R_3 and R_3/R_2.

The voltage gain, before feedback has been applied, of the transistor circuit of Fig. 4.8a is $A_v = A_i R_3/R_{IN}$ and thus the gain with feedback is

$$A_{v(F)} = \frac{A_i R_3/R_{IN}}{1+\dfrac{R_4}{R_3}\cdot\dfrac{A_i R_3}{R_{IN}}} = \frac{A_i R_3}{R_{IN} + A_i R_4} \qquad (4.5)$$

Similarly, the voltage gain of the f.e.t. circuit shown in Fig. 4.8b is

$$A_{v(F)} = \frac{g_m R_2}{1+\dfrac{R_3}{R_2}g_m R_2} = \frac{g_m R_2}{1+g_m R_3} \qquad (4.6)$$

EXAMPLE 4.5

An amplifier of the type shown in Fig. 4.8b has the following data: $R_2 = 4.7$ kΩ, $R_3 = 1$ kΩ and $g_m = 5$ mS. Calculate the voltage gain of the circuit.

Solution
From equation (4.6)

$$A_{v(F)} = \frac{5\times10^{-3}\times4.7\times10^3}{1+5\times10^{-3}\times1\times10^3} = 3.92 \qquad (Ans.)$$

Current-Current Feedback

Current-current feedback involves feeding a fraction of the output current of an amplifier into the input circuit in antiphase with the input signal current (Fig. 4.9). The current amplifier is assumed to have a very low input impedance and

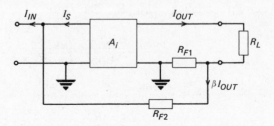

Fig. 4.9 Current–current negative feedback

the source of the signal current I_{IN} is assumed to have a very high impedance. The output current I_{OUT} passes through the load and then splits into two parts: one part flows via R_{F1} to earth and the other part flows via R_{F2} and the input impedance of the amplifier to earth. Assuming that R_{F2} is much greater than the input impedance of the amplifier, the feedback current βI_{OUT} is

$$\beta I_{OUT} = \frac{I_{OUT} R_{F1}}{R_{F1} + R_{F2}}$$

so that

$$\beta = \frac{R_{F1}}{R_{F1} + R_{F2}} \tag{4.7}$$

The fed-back current βI_{OUT} is in antiphase with the input current I_{IN}; therefore

$$I_S = I_{IN} - \beta I_{OUT} \quad \text{and} \quad I_{OUT} = A_i I_S = A_i (I_{IN} - \beta I_{OUT})$$
$$I_{OUT}(1 + \beta A_i) = A_i I_{IN}$$
$$I_{OUT} = \frac{A_i I_{IN}}{1 + \beta A_i}$$

The current gain with feedback is therefore

$$A_{i(F)} = \frac{I_{OUT}}{I_{IN}} = \frac{A_i}{1 + \beta A_i} \tag{4.8}$$

Fig. 4.10 shows the circuit of a two-stage amplifier with current-current feedback applied by connecting a feedback resistor R_7 between the emitter of the second stage and the base of the first stage.

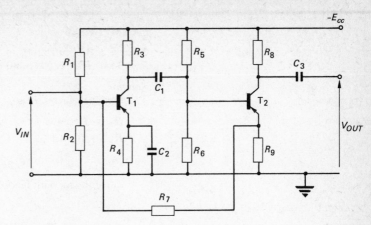

Fig. 4.10 A two-stage current–current n.f.b. amplifier

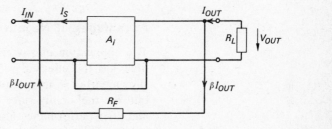

Fig. 4.11 Current-voltage negative feedback

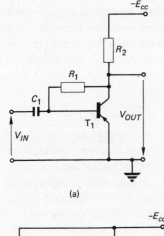

(a)

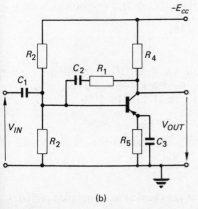

(b)

Fig. 4.12 Two current-voltage negative-feedback circuits

Current-Voltage Feedback

The fourth type of feedback involves feeding back to the input in antiphase with the input signal current, a current whose magnitude is proportional to the output voltage. The basic arrangement of *current-voltage* feedback is shown in Fig. 4.11.

Provided that the feedback resistance R_F is much larger than input impedance of the amplifier, $\beta I_{OUT} \approx V_{OUT}/R_F$. The output current I_{OUT} is equal to V_{OUT}/R_L, and therefore

$$\frac{\beta I_{OUT}}{I_{OUT}} = \frac{V_{OUT}/R_F}{V_{OUT}/R_L}$$

so that

$$\beta = \frac{R_L}{R_F} \tag{4.9}$$

The simplest current-voltage feedback circuit is shown in Fig. 4.12a; the feedback current βI_{OUT} is applied to the input terminals via the feedback resistor R_1 connected between the collector and base terminals. R_1 also provides d.c. bias and stabilization for the circuit. Sometimes the value of R_1 required to provide a desired amount of feedback is not the same as that required for correct bias; then a capacitor is connected in series with R_1 and the bias is provided by some other means, such as that shown in Fig. 4.12b.

EXAMPLE 4.6

Calculate the current gain of the circuit shown in Fig. 4.12a if $R_2 = 3\,\mathrm{k\Omega}$, $R_1 = 120\,\mathrm{k\Omega}$ and the current gain of the transistor is 120.

Solution
The feedback factor β is

$$\frac{R_2}{R_1} = \frac{3}{120} = \frac{1}{40}$$

Therefore

$$\text{Current gain with feedback} = A_{i(F)} = \frac{120}{1 + 120 \times \dfrac{1}{40}} = \frac{120}{4}$$
$$= 30 \qquad (Ans.)$$

Advantages of Negative Feedback

The application of any type of negative feedback to an amplifier has the effect of reducing the gain of that amplifier, as shown earlier in this chapter. This is obviously a disadvantage, but on the credit side, a number of desirable changes in amplifier performance also occur, namely the gain stability is considerably increased, distortion and noise produced within the feedback loop are reduced, and the input and output impedances of the amplifier can be modified to almost any desired value.

Stability of Gain

For many applications it is important that an amplifier should have a constant gain even though various parameters, for example power supply voltage and transistor current gain, may alter with time and/or change in ambient temperature. The application of negative feedback will considerably reduce the effect of such parameter variations on the overall gain of the amplifier.

EXAMPLE 4.7

An amplifier has a voltage gain of 1000. If 3/100 of the output voltage is applied as negative feedback, calculate the change in overall gain if the gain before feedback falls by 50%.

Solution
Gain with feedback, inherent gain being 1000,

$$A_{v(F)1} = \frac{1000}{1 + (1000 \times 3/100)} = \frac{1000}{31} \approx 32.6$$

New gain with feedback, inherent gain having been reduced to 500,

$$A_{v(F)2} = \frac{500}{1 + (500 \times 3/100)} = \frac{500}{16} = 31.25$$

Therefore

$$\text{Change in gain} = \frac{32.26 - 31.25}{32.26} \times 100 = -3.13\% \qquad (Ans.)$$

Thus a 50% fall in the inherent gain of the amplifier results in only a 3.13% change in the overall gain.

If the loop gain, βA_v or βA_i, is much larger than unity then equations (4.1) and (4.8) become

$$A_{v(F)} \approx \frac{A_v}{\beta A_v} = \frac{1}{\beta} \qquad (4.10)$$

$$A_{i(F)} \approx \frac{A_i}{\beta A_i} = \frac{1}{\beta} \qquad (4.11)$$

It is evident that, if $\beta A \gg 1$, the gain of the amplifier is merely a function of the feedback circuit and is quite independent of the characteristics of the amplifier itself. Any changes in the performance of the amplifier will not now affect the overall gain. If the feedback network is purely resistive, the amplifier will have a uniform gain/frequency characteristic at all frequencies at which the loop gain βA remains much larger than unity (Fig. 4.13). Should a non-uniform gain/frequency characteristic be required a feedback network containing capacitive components is required (Fig. 4.13b,c,d). The resistance-capacitance feedback network is chosen to have a loss/frequency characteristic that is the inverse of the required amplifier gain/frequency characteristic.

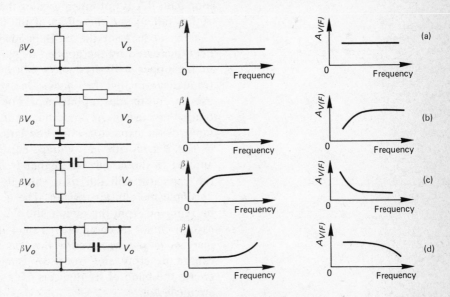

Fig. 4.13 Showing various β networks and the corresponding amplifier gain/frequency characteristics if $\beta A_v \gg 1$

Amplitude/Frequency Distortion

Amplitude/frequency distortion of a complex signal is caused by the different frequency components of the signal being amplified, or attenuated, to different extents. In an amplifier this kind of distortion is produced if the gain of the amplifier is not the same at all frequencies; for example, the gain of an R-C coupled amplifier falls off at both high and low frequencies because of capacitive effects. The application of negative

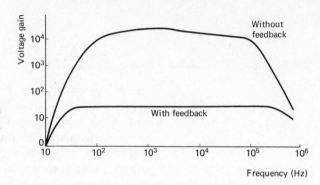

Fig. 4.14 Showing the effect of negative feedback on the gain/frequency characteristics of an amplifier

feedback to an amplifier reduces change in its gain and thereby reduces amplitude/frequency distortion. Fig. 4.14 shows the gain/frequency characteristic of a typical amplifier before and after the application of negative feedback. Clearly the feedback has made the amplifier gain much "flatter" over most of the frequency band. At high and low frequencies, however, the loop gain βA is not much greater than unity (because the gain A has fallen) and the effect of the feedback is reduced.

This means that the 3 dB bandwidth of the amplifier has been increased by the applied negative feedback. An increase in the upper 3 dB frequency can be obtained by the use of frequency-sensitive feedback. One way of achieving this is to use an emitter decoupling capacitor of much smaller capacitance than usual. At low and medium frequencies the reactance of the capacitor will be so large that the emitter resistor is not decoupled and voltage-current negative feedback is applied to the circuit. At higher frequencies the reactance of the capacitor will fall to a value low enough to adequately decouple the emitter resistor. The negative feedback will then be removed from the circuit and the voltage gain will rise. By careful choice of component values it can be arranged for the increase in gain achieved in this way to balance the fall in gain caused by stray and transistor capacitances and thereby increase the band of frequencies over which the response of the circuit is flat.

Non-linearity Distortion

The dynamic transfer and mutual characteristic of all transistors and f.e.t.s exhibit some non-linearity, and as a result the output signal waveform is not identical with that of the input signal. *Non-linearity distortion* is said to have occurred. Analysis beyond the scope of this book shows that the output waveform is distorted because it contains components at frequencies that were not present in the input signal. If these extra, unwanted frequencies are harmonically related to the frequency of the input signal HARMONIC DISTORTION is said to be present. If the new frequencies are equal to the sums and differences of frequencies contained in the input signal, or harmonics of them, INTERMODULATION DISTORTION has occurred. For example, suppose the input signal contains frequencies f_1 and f_2. If the output signal has components at $2f_1$, $2f_2$, $3f_1$, $3f_2$, etc., harmonic distortion has occurred; if components at $2f_1 \pm f_2$, $2f_2 \pm f_1$, etc., exist intermodulation distortion is present. Often both harmonic and intermodulation distortion occur at the same time.

The line amplifiers in a multi-channel telephony system handle the composite signal produced by a number of channels simultaneously passing signals. If non-linearity is present in the line amplifiers, the many intermodulation components produced cause crosstalk between the various channels which manifests itself at the output of each channel as INTERMODULATION NOISE. To reduce intermodulation noise to a tolerable level the line amplifiers must have the minimum practical non-linearity, and to this end considerable negative feedback is applied.

It can be shown that non-linearity distortion in an amplifier is reduced by the application of negative feedback according to the following equation:

$$\text{Distortion with n.f.b. applied} = \frac{\text{Distortion without n.f.b.}}{1 + \beta A}$$

$$(4.12)$$

assuming the same output signal level in each case. It can be seen that the distortion is reduced to the same extent as is the gain of the amplifier and it might seem that little has been achieved. However, non-linearity distortion occurs mainly in the final, large-signal stage of an amplifier, and if it can be reduced by negative feedback the lost gain can be made up in earlier stages and an overall reduction in distortion obtained.

EXAMPLE 4.8

An amplifier has a voltage gain of 50 dB. Find the change in the gain if 1/50 of the output voltage is fed back into the input in opposition to the input signal. What is then the reduction in harmonic distortion at the output of the amplifier? Phase shift in the amplifier and the feedback path may be neglected.

Solution
Equation (4.1) is

$$A_{v(F)} = \frac{A_v}{1 + \beta A_v}$$

A_v is the voltage gain of the amplifier before feedback has been applied, expressed as a voltage ratio, not in decibels. The first step, therefore, is to convert the given gain of 50 dB into the corresponding voltage ratio. Since no information about the input and load impedances of the amplifier is given, we must assume that these impedances are equal; this allows the expression dB = 20 log$_{10}$ (voltage ratio) to be used:

50 = 20 log$_{10}$ (voltage ratio)

50/20 = 2.5 = log$_{10}$ (voltage ratio)

Taking antilog$_{10}$ of sides

316.2 = voltage ratio

Therefore

$$A_{v(F)} = \frac{316.2}{1 + (316.2/50)} = 43.2 \text{ times} = 32.7 \text{ dB}$$

so that

Change in gain = 50 − 32.7 = 17.3 dB   (*Ans.*)

From equation (4.12), the reduction in harmonic distortion is

$$\frac{\text{Distortion with feedback}}{\text{Distortion without feedback}} = \frac{1}{1 + (316.2/50)}$$

$$= \frac{1}{7.32} \approx 0.137 \text{ times} \quad (Ans.)$$

Noise

The various sources of noise in amplifiers will be discussed in Chapter 9; here noise may be considered as being any unwanted sound. The application of negative feedback to an amplifier reduces any noise produced or picked up *within the feedback loop* according to the equation

$$\text{Noise with n.f.b. applied} = \frac{\text{Noise without n.f.b.}}{1 + \beta A} \quad (4.13)$$

Input and Output Impedances

The application of negative feedback to an amplifier alters its input and output impedances. Each impedance may be either increased or decreased depending on the type of feedback used; see Table 4.1.

Table 4.1

Type of n.f.b.	Input impedance	Output impedance
Voltage-voltage	Increased	Decreased
Voltage-current	Increased	Increased
Current-current	Decreased	Increased
Current-voltage	Decreased	Decreased

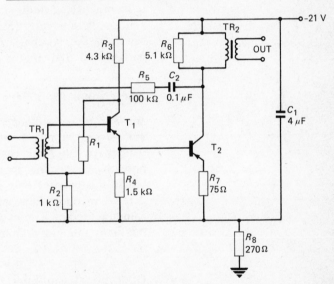

Fig. 4.15 The standard Post Office audio-frequency line amplifier (*Post Office Electrical Engineer's Journal*)

The standard B.P.O. amplifier (Fig. 4.15) for four-wire audio-frequency telephone circuits has been designed to work between 600 Ω impedances, and to have a maximum gain of 30 dB and a maximum output level of 50 mW, i.e. +17 dBm.† The amplifier is normally set up to have a maximum output level of +10 dBm, the lower level resulting in a reduction in the distortion content of the output signal.

Resistors R_1, R_2, R_3 and R_4 provide the bias for the first stage. The first stage is operated as an emitter follower and therefore has a high input impedance and a low output impedance. The second stage is a common-emitter amplifier with voltage-current negative feedback applied because the emitter resistor R_7 is not decoupled. This feedback increases both the input and output impedances of the stage. Direct coupling between the stages is employed and overall negative feedback is applied via R_5 and C_2.

† dBm—decibels relative to 1 mW.

Instability in Negative Feedback Amplifiers

At most signal frequencies there is 180° phase shift through each stage of a multi-stage amplifier and the feedback voltage is arranged to be in antiphase with the input signal voltage. At both low and high frequencies the phase shift through the amplifier will be altered because of the inevitable capacitive effects. If at a particular high frequency the total phase shift due to the stray and transistor capacitances is equal to 180°, the feedback will become positive. There is then the possibility of the amplifier oscillating unless the loop gain βA_v at this frequency is less than unity. There is also a low frequency, at which the overall shift caused by the coupling capacitors is 180°, where oscillation is equally possible.

In practice, a multi-stage amplifier is *compensated* to prevent oscillation; this consists of the addition of suitable capacitance to the feedback network to so shape the frequency variation of βA_V that oscillation is impossible.

Operational Amplifiers

An *operational amplifier* is a circuit that has a very high voltage gain, a high input impedance, and a low output impedance. An OP-AMP can be used for a wide variety of purposes by connecting appropriate circuitry across its terminals but in this chapter only its use as an amplifier will be considered. An operational amplifier can be made using discrete components but it is nowadays more common to use one of the many integrated circuit versions available.

Operational Amplifier Parameters

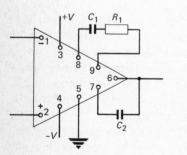

Fig. 4.16 The operational amplifier

The symbol for an operational amplifier is shown in Fig. 4.16. Two input terminals, 1 and 2, are provided. One of them, labelled −, is known as the inverting input terminal since a signal applied to this terminal appears at the output terminal 6 with the opposite polarity, i.e. a sinusoidal input signal will experience a phase shift of 180°. The other input terminal, labelled +, is the non-inverting input terminal, and a signal applied to this terminal is amplified with zero phase shift. Two further terminals, 3 and 4, are provided for the connection of positive and negative power supply voltages. Most op-amps will operate satisfactorily with a wide range of supply voltages although it is generally advisable to decouple each power supply line with a suitable value of capacitance. Another terminal, 5, is provided to which the mid-point of the power supplies must be connected; in most cases this point is earth potential as shown in the diagram. The output voltage appears at terminal 6 and is developed as a potential relative to earth. Further terminals, 7, 8 and 9, are provided, across which components can be connected to ensure the stability of the amplifier under all conditions. Not all types of op-amp require external *compensation* components to be fitted.

(*a*) AMPLIFIER GAIN
The ideal operational amplifier would have an infinite open-loop gain but, naturally, practical circuits fall far short of this. Commercially available op-amps have open-loop gains which vary considerably from one type to another but may be somewhere between 10 000 and 200 000.

(*b*) INPUT IMPEDANCE
Ideally, the input impedance of an op-amp is infinitely high but in practice may vary from $250 \, k\Omega$ to $2 \, M\Omega$ for bipolar input types and perhaps $10^{12} \, \Omega$ for a f.e.t. input type.

(*c*) OUTPUT IMPEDANCE
Since an op-amp is essentially a voltage amplifier its output impedance should be as low as possible. Practical output impedances are in the region of $150 \, \Omega$.

(*d*) COMMON-MODE REJECTION RATIO
The output voltage of an operational amplifier is proportional to the difference between the voltages applied to its inverting and non-inverting terminals. When the two voltages are equal, the output voltage should be zero. A signal applied to both input terminals is known as a *common-mode signal* and is nearly always an unwanted noise voltage. The ability of an operational amplifier to suppress common-mode signals is expressed in terms of its *common-mode rejection ratio* (c.m.r.r.). The c.m.r.r. is defined by equation (4.14), i.e.

$$\text{Common-mode rejection ratio (c.m.r.r.)} = 20 \log_{10}$$

$$\times \frac{\text{Voltage gain for signal applied to + or - terminal}}{\text{Common-mode gain}} \, dB$$

$$(4.14)$$

Typically a commercial op-amp might have a c.m.r.r. of 80 dB.

(*e*) SLEW RATE
The slew rate of an operational amplifier is the maximum rate, in volts/μsecond, at which its output voltage is capable of changing when the maximum output voltage is being supplied. When a signal at a given frequency is applied to an op-amp the maximum permissible output voltage is determined by the slew

rate; should a greater output voltage be developed the signal waveform will be distorted. This means that an op-amp can be used to provide either a large output voltage or a high upper 3 dB frequency, but not both at the same time.

Typical slew rates are in the region of $1-2$ V/μs.

(f) BANDWIDTH

The open-loop voltage gain of an op-amp is not constant at all frequencies but falls at high frequencies because of capacitive effects. The gain/frequency characteristic of an op-amp can be specified by the manufacturer in more than one way. These are: (i) the upper 3 dB frequency can be quoted, (ii) the bandwidth over which the gain/frequency characteristic is flat can be quoted; and (iii) the frequency at which the gain falls to unity can be quoted; this is analogous to the f_t of a bipolar transistor and is possible since at frequencies above the 3 dB frequency the gain falls at the constant rate of 6 dB/octave.

(g) INPUT OFFSET VOLTAGE

Ideally, the output voltage of an operational amplifier should be zero when zero voltage is applied to both its input terminals. For any practical amplifier it is found that an output voltage does exist for zero input voltage. This voltage arises because of imperfections within the amplifier but, for convenience, is presumed to be caused by an input *offset voltage*. The input offset voltage of an operational amplifier is equal to the output voltage for zero input voltage divided by the open-loop voltage gain of the amplifier. Typically, the input offset voltage is about 1 mV.

(h) INPUT OFFSET CURRENT

An operational amplifier will have small bias currents flowing into its two input terminals even when the input voltages are zero. The difference between these bias currents is known as the *input offset current*.

Inverting and Non-inverting Amplifiers

When an operational amplifier is used as an amplifying element, a large amount of negative feedback is applied to specify accurately the voltage gain. There are two ways in which an operational amplifier can be connected to act as an amplifier; one method gives an inverting gain and the other gives a non-inverting gain. The connections for an inverting gain are shown in Fig. 4.17a. A resistor R_1 is connected between the input terminal of the amplifier circuit and the inverting input terminal of the op-amp, and another resistor R_2 is connected between the inverting input terminal and the output terminal.

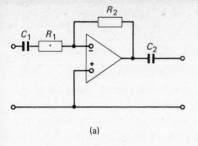

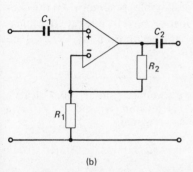

(a)

(b)

Fig. 4.17 Operational amplifier connected to provide (a) an inverting gain and (b) a non-inverting gain

The non-inverting terminal is connected to earth. The open-loop gain of an operational amplifier is very high and this means that the voltage at the inverting terminal must be very small. The input voltage will be very nearly at the same potential as the non-inverting terminal and is said to be a *virtual earth*. The input voltage V_{IN} appears across R_1 and the input current is V_{IN}/R_1. The input impedance of the op-amp is high and so very little current flows into the op-amp itself. All the input current therefore flows through the resistor R_2 and the voltage developed across R_2 is equal to the output voltage V_{OUT} of the circuit. Therefore

$$V_{IN}/R_1 = -V_{OUT}/R_2$$

or

$$A_{v(F)} = \frac{V_{OUT}}{V_{IN}} = \frac{-R_2}{R_1} \tag{4.15}$$

Current-voltage negative feedback has been applied to the op-amp and so both its input and output impedances are reduced to a low value. The input resistance of the op-amp appears in series with the resistor R_1 and so the input resistance of the circuit is approximately R_1 ohms.

In the case of the non-inverting amplifier, shown in Fig. 4.17b, the output voltage V_{OUT} of the circuit is equal to $A_v(V_{IN} - V_x)$, where A_v is the open-loop gain of the op-amp and V_x is the voltage which appears at the inverting terminal, i.e.

$$V_x = \frac{V_{OUT}R_1}{R_1 + R_2}$$

Therefore

$$V_{OUT} = A_v\left[V_{IN} - \frac{V_{OUT}R_1}{R_1 + R_2}\right]$$

$$V_{OUT}\left[1 + \frac{A_v R_1}{R_1 + R_2}\right] = A_v V_{IN}$$

and

$$A_{v(F)} = \frac{V_{OUT}}{V_{IN}} = \frac{A_v}{1 + \dfrac{A_v R_1}{R_1 + R_2}}$$

The inherent voltage gain A_v of the operational amplifier is very large and so $A_v R_1/(R_1 + R_2) \gg 1$ and therefore

$$A_{v(F)} = \frac{A_v}{A_v R_1/(R_1 + R_2)}$$

or

$$A_{v(F)} = \frac{R_1 + R_2}{R_1} \qquad (4.16)$$

The non-inverting amplifier is constructed by applying voltage-voltage negative feedback to the operational amplifier. The input resistance of the operational amplifier, and hence of the non-inverting amplifier, is increased to a very high value, while its output resistance is reduced to a low value.

In Figs. 4.17a and b the input and output capacitors act as d.c. blocks and should have negligible reactance at the frequencies of interest. For very low or zero frequency amplification, of course, the capacitors will be omitted.

EXAMPLE 4.9

An operational amplifier is to have a voltage gain of 100. Calculate the required values for the external resistors R_1 and R_2 if: (a) a non-inverting, (b) an inverting gain is required.

Solution
(a) From equation (4.16)

$$100 = \frac{R_1 + R_2}{R_1} \quad \text{or} \quad 99\,R_1 = R_2$$

Hence, if R_2 is chosen to be $1\,\text{M}\Omega$, R_1 should be equal to

$$\frac{1 \times 10^6}{99} = 10.1\,\text{k}\Omega \qquad (Ans.)$$

(b) From equation (4.15)

$$-100 = -\frac{R_2}{R_1}$$

Choosing R_2 as $1\,\text{M}\Omega$, gives

$$R_1 = \frac{1 \times 10^6}{100} = 10\,\text{k}\Omega \qquad (Ans.)$$

The operational amplifier equivalent of the emitter or source follower is known as the VOLTAGE FOLLOWER and its circuit is shown in Fig. 4.18. A direct connection is made between the inverting input and output terminals and therefore 100% voltage-voltage negative feedback is applied. The circuit has a very high input impedance, a very low output impedance, and a voltage gain that is very nearly equal to unity. The main application of the voltage follower is as a means of connecting a high-impedance source to a low-impedance load, i.e. to act as a BUFFER amplifier.

Fig. 4.19 shows how an operational amplifier can be employed to add two or more voltages together. The circuit is

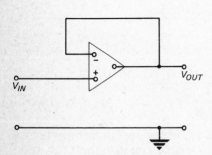

Fig. 4.18 The voltage follower

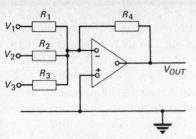

Fig. 4.19 Operational adder

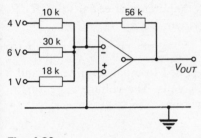

Fig. 4.20

essentially an inverting amplifier with more than one input terminal and hence

$$V_1/R_1 + V_2/R_2 + V_3/R_3 = -V_{OUT}/R_4$$

or

$$V_{OUT} = -R_4[V_1/R_1 + V_2/R_2 + V_3/R_3] \qquad (4.17)$$

If $R_1 = R_2 = R_3 = R_4$ then this equation reduces to

$$V_{OUT} = -(V_1 + V_2 + V_3) \qquad (4.18)$$

EXAMPLE 4.10

Determine the output voltage of the circuit shown in Fig. 4.20.

Solution
From equation (4.17)

$$V_{OUT} = -56[\tfrac{4}{10} + \tfrac{6}{30} + \tfrac{1}{18}] = -36.71\,\text{V} \qquad (Ans.)$$

Frequency Compensation

An operational amplifier is a multi-stage circuit having a very high open-loop voltage gain. Either voltage-voltage or current-voltage negative feedback is applied to the amplifier to specify a very much lower voltage gain.

At higher frequencies the open-loop gain of an op-amp falls because of internal capacitive effects and these also produce internal phase shifts additional to those present at lower frequencies between the inverting input and the output terminals. An amplifier using an op-amp as the active device may therefore become unstable and perhaps oscillate at a high frequency. To ensure the stability of an op-amp circuit it is necessary to apply *frequency compensation* to the device. Some types of op-amp are internally compensated and are stable for any amount of applied negative feedback but such amplifiers generally have a limited bandwidth. Other types of op-amp are not provided with internal compensation and then require the connection of suitable compensating components across the appropriate terminals. Generally, a series resistance-capacitance network is employed but sometimes a capacitor alone is sufficient. Also, some circuits require both a series R-C circuit connected between one pair of terminals, and a capacitor connected between a third terminal and the output terminal. An example of this is shown in Fig. 4.16. When external frequency compensation is used, the component values can be chosen to give maximum bandwidth for the amount of negative feedback which is applied. Typically, $C_1 = 5\,\text{nF}$, $R_1 = 1.5\,\text{k}\Omega$ and $C_2 = 200\,\text{pF}$ or $C_1 = 10\,\text{pF}$, $R_1 = 0$ and $C_2 = 3\,\text{pF}$.

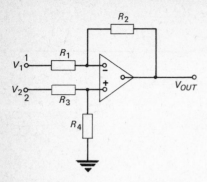

Fig. 4.21 Differential amplifier

Differential Amplifier

When an operational amplifier is used as a differential amplifier (Fig. 4.21) voltages are applied to its two input terminals, 1 and 2, and the *difference* between these voltages is amplified.

Suppose that voltage V_1 volts is applied to terminal 1 and zero volts to terminal 2. The difference in the potentials at the inverting and non-inverting op-amp terminals is very nearly zero and therefore the inverting terminal must be at zero potential. This means that the input voltage V_1 is developed across resistor R_1 and the input current is $I_1 = V_1/R_1$. Since the input impedance of the op-amp is high this current flows through resistor R_2. The voltage dropped across R_2, which is the output voltage V_{OUT} of the circuit, is equal to $V_1 R_2/R_1$ and the voltage gain of the circuit is

$$A_v = V_{OUT}/V_1 = -R_2/R_1 \qquad (4.19)$$

Conversely, if the voltages applied to input terminals 1 and 2 are, respectively, zero and V_2 volts. The voltage appearing at the non-inverting terminal will be

$$V_2 R_4/(R_3 + R_4) \text{ volts}$$

This voltage will also appear at the inverting terminal and so the voltage across resistor R_1 must also be equal to

$$-V_2 R_4/(R_3 + R_4)$$

The output voltage V_{OUT} of the circuit is now

$$V_{OUT} = V_2 R_4/(R_3 + R_4) + [-V_2 R_4/(R_3 + R_4)] \times -R_2/R_1$$

and the voltage gain A_v of the circuit is

$$A_v = V_{OUT}/V_2 = [R_4/(R_3 + R_4)][1 + R_2/R_1] \qquad (4:20)$$

Lastly, if the voltages applied to terminals 1 and 2 are V_1 volts and V_2 volts respectively the difference between the two voltages is amplified. If $V_1 > V_2$ a voltage $(V_1 - V_2)$ is amplified $-R_2/R_1$ times, whilst if $V_2 > V_1$ a voltage $(V_2 - V_1)$ is amplified by

$$[R_4/(R_3 + R_4)][1 + R_2/R_1]$$

EXAMPLE 4.11

In a differential amplifier of the type shown in Fig. 4.21, $R_1 = 10 \text{ k}\Omega$ $R_2 = 100 \text{ k}\Omega$, $R_3 = 10 \text{ k}\Omega$ and $R_4 = 100 \text{ k}\Omega$. Calculate the output voltage of the circuit if (i) $V_1 = 10 \text{ mV}$, $V_2 = 0$, (ii) $V_1 = 0$, $V_2 = 10 \text{ mV}$, (iii) $V_1 = 100 \text{ mV}$, $V_2 = 50 \text{ mV}$, and (iv) $V_1 = 50 \text{ mV}$, $V_2 = 100 \text{ mV}$.

Solution

(i) From equation (4.19)

$$V_{OUT} = \frac{-100}{10} \times 10 \text{ mV} = -100 \text{ mV} \qquad (Ans.)$$

(ii) From equation (4.20),

$$V_{OUT} = \frac{100}{110} \left(1 + \frac{100}{10} \right) \times 10 \text{ mV} = +100 \text{ mV} \qquad (Ans.)$$

(iii) V_1 is larger than V_2 and hence

$$V_{OUT} = \frac{-100}{10} \times 50 \text{ mV} = -500 \text{ mV} \qquad (Ans.)$$

(iv) V_2 is 50 mV larger than V_1 and hence

$$V_{OUT} = \frac{100}{110} \left(1 + \frac{100}{10} \right) \times 50 \text{ mV} = +500 \text{ mV} \qquad (Ans.)$$

Practical Operational Amplifiers

A wide variety of integrated circuit operational amplifiers are available from a number of manufacturers. Many types of operational amplifiers have become standard types and are manufactured by more than one firm. Manufacturers identify operational amplifiers by means of their own code numbers but the numbering of a particular type includes the same three numbers regardless of manufacturer.

For example, one of the most popular op-amps in use today is the 741.

This device is labelled in the following ways by the manufacturers quoted:

Signetics   A 741CV   Fairchild   A 741TC
Motorola   MC 1741CP1
RCA   CA 741CS   Texas   SN 2741P

The characteristics of the 741 operational amplifier are open-loop voltage gain 200 000, input impedance 2 MΩ, output impedance 75 Ω, slew rate 0.5 V/μs, common-mode rejection ratio 90 dB, bandwidth for unity gain 1 MHz. The circuit is internally compensated and has internal protection against damage caused by output short-circuits or high input voltages

The 747 operational amplifier is simply two 741 op-amps in the one package. The 748 circuit has the same characteristics as the 741 but it is not internally compensated against instability. The external compensation required is generally only a capacitor connected between the two appropriate terminals and results in a wider gain bandwidth product than the 741 can provide.

One of the earliest operational amplifiers to be offered for sale was the 709 and it is still sold by most manufacturers today. The 709 has an open-loop voltage gain of 45 000 and input impedance of 250 kΩ, an output impedance of 150 Ω, a common-mode rejection ratio of 90 dB, a slew rate of 0.3 V/μs and a unity gain bandwidth of 1 MHz.

Exercises

4.1. Draw the circuit diagram and explain the operation of a two-stage transistor amplifier for use as a repeater in an audio transmission system. Explain the precautions taken for stabilizing against temperature variations. (C & G)

4.2. Draw a circuit diagram of an audio-frequency transistor amplifier comprising a common-emitter stage in tandem with an emitter-follower stage. Show clearly the biasing arrangements and briefly explain their operation. State the order of magnitude of the input and output impedances of the amplifier. Say what factors will affect the low-frequency response of the circuit you have drawn. (C & G)

4.3. Explain briefly why negative feedback is used on carrier line amplifiers.

An amplifier has a gain of 60 dB without feedback and 30 dB when feedback is applied. If the gain without feedback changes to 55 dB calculate the new gain with feedback. (C & G)

4.4. Draw a circuit diagram and describe the operation of a simple two-stage low-frequency amplifier using transistors, to operate with an input resistance of about $800\,\Omega$ and with an output resistance of about $100\,\Omega$.

Explain the term *thermal runaway* and describe a method of preventing it. (C & G)

4.5. A fraction β of the output voltage of an amplifier is fed back in antiphase with the input voltage. Derive an expression giving the resultant gain of the amplifier as a function of the gain without feedback and the fraction β.

The maximum voltage gain of an amplifier without feedback is 1000 times and the variation of gain over its working range of frequencies is 4 dB. Calculate the resultant variation of gain in decibels if 0.1 of the output voltage is fed back, in antiphase, to the input. Neglect any possible effects of phase variations.

What are the principal advantages of using negative feedback in coaxial line amplifiers? (C & G)

4.6. With the aid of a circuit diagram, describe the operation of the emitter follower. Determine an expression for the stage gain using the general feedback formula. In the circuit chosen what effect has feedback on the following: (*a*) input impedance, (*b*) output impedance, (*c*) signal distortion?

Describe briefly one application of the circuit chosen.
(C & G)

4.7. Sketch the circuit diagram of an emitter follower. Describe the properties of the circuit and explain why it is a voltage feedback system. An emitter follower has an a.c. input resistance of $160\,k\Omega$ and a current gain of 70. The load resistor has a resistance of $2.2\,k\Omega$. Determine the power gain in decibels. State where this type of circuit can be used in one type of instrument. (C & G)

4.8. (*a*) Explain what is meant by the following types of distortion in the output of an amplifier: (i) harmonic, (ii) amplitude/frequency, (iii) phase/frequency.

(*b*) In each of the above cases explain a possible cause and state a method of reducing the distortion. (C & G)

4.9. An amplifier has a voltage gain of 300 and input and load resistances of $5\,k\Omega$ and $1\,k\Omega$ respectively. When negative feedback is applied the input resistance is $100\,k\Omega$ and the voltage gain is 15. (*a*) Determine the power gain in dB (i) without negative feedback, (ii) with negative feedback.

(*b*) What is the value of the feedback factor? (C & G)

4.10. (*a*) The voltage gain of an amplifier is *A*. If a fraction *B* of the output voltage is fed back in anti-phase with the input, derive the modified gain of the amplifier. (*b*) What is the value of *B* in an emitter follower and the approximate voltage gain for this device? (*c*) Draw the basic circuit of an emitter follower showing bias arrangements and input and output connections. (*d*) Give two applications for an emitter follower.   (C & G)

4.11. (*a*) Describe with the aid of a circuit diagram, the operation of either a source follower or a voltage follower. (*b*) What effect has the negative feedback on the gain, and the input and output impedances of the circuit? (*c*) Describe a typical application for your circuit.

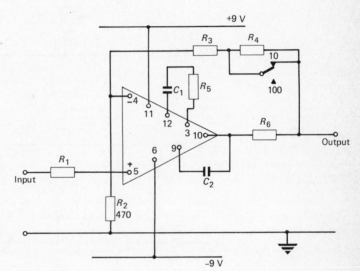

Fig. 4.22

4.12. Fig. 4.22 shows a circuit diagram for a low-frequency voltage amplifier which can be switched to give a precise gain of either ten or one hundred.
(*a*) Calculate the values of the two resistors R_3 and R_4.
(*b*) Give approximate values for other external components in the circuit.
(*c*) Explain briefly the purpose of (i) C_1 and R_5, (ii) C_2, (iii) R_6.
          (C & G)

4.13. (*a*) With reference to Fig. 4.23 state why (i) R_5/R_6 is de-coupled, (ii) R_7 is not decoupled (iii) R_3 is returned to R_5/R_6.
(*b*) Is the input impedance raised, lowered, or unchanged by the feedback caused by (i) R_2, (ii) the R_3/R_7 combination?
          (C & G)

Short Exercises

4.14. List the four types of negative feedback and state the effect of each on the output and input impedances of an amplifier.

4.15. An amplifier is to have negative feedback applied to make it approach the ideal for (i) a voltage and (ii) a current amplifier. Which type of feedback should be used in each case?

4.16. Draw diagrams to show (i) an emitter follower, (ii) a source follower, and (iii) a voltage follower. What is the main application of these circuits?

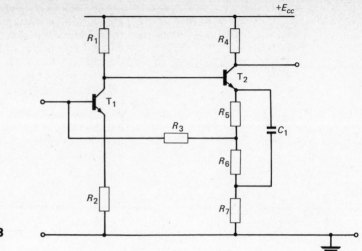

Fig. 4.23

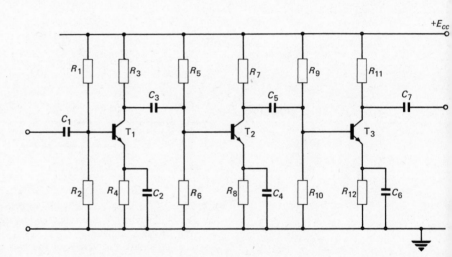

Fig. 4.24

Table A

Effect of n.f.b. on:	UP	DOWN
gain		\
gain stability	\	
bandwidth		\
distortion		\
noise		\
input impedance		\
output impedance	\	

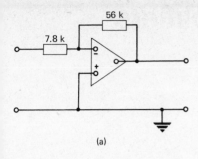

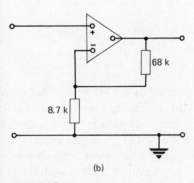

(a)

(b)

Fig. 4.25

4.17. Fig. 4.24 shows the circuit diagram of a three-stage *R-C* coupled amplifier. How would you apply negative feedback to the circuit? What type of n.f.b. have you drawn?

4.18. Complete Table A by ticking the appropriate columns.

4.19. What are the advantages of integrated circuit operational amplifiers over the discrete component versions?

4.20. Calculate the voltage gain of the circuits shown in Figs. 4.25*a* and *b*.

4.21. Calculate the voltage gain of the circuit shown in Fig. 4.26.

4.22. Explain the meanings of the terms slew rate and common-mode rejection ratio as applied to an operational amplifier.

4.23. List the function of each component shown in Fig. 4.27.

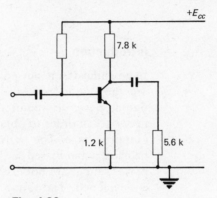

Fig. 4.26

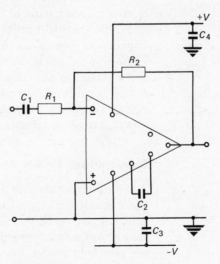

Fig. 4.27

5 Audio-frequency Power Amplifiers

Introduction

In an audio-frequency power amplifier the main considerations are the output power and the efficiency. If an appreciable output power is required, a large amplitude input signal is necessary in order to obtain large swings of output current and output voltage. The transistor used in a power amplifier must be chosen and biased so that its maximum current, voltage and power ratings are not exceeded when the desired output power is developed. The rating figures of a particular transistor are available from manufacturers data sheets.

For the maximum power to be delivered by a power amplifier to its load, without exceeding a predetermined distortion level, the transistor must work into a particular value of load impedance, known as the OPTIMUM LOAD. Power amplifiers using discrete components are either single-ended, i.e. use a single output transistor, or employ two transistors connected in push-pull. In addition, a number of integrated circuit power amplifiers, capable of providing several watts output power, are also available.

Single-ended Power Amplifiers

The output transistor of a single-ended power amplifier should work into its optimum load impedance. Rarely will the actual load impedance be equal to this optimum value and so transformer coupling is normally used, Fig. 5.1. Transformer coupling also reduces the d.c. power lost in the circuit because the transformer primary winding usually has a low d.c. resistance.

The turns ratio, n $(=N_1/N_2)$, of the output transformer is chosen to transform the actual load R_L into the optimum load R_L' of the transistor, i.e.

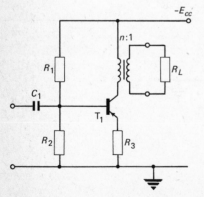

Fig. 5.1 Single-ended power amplifier

$$n = \sqrt{\frac{R_L'}{R_L}}$$

(5.1)

EXAMPLE 5.1

The optimum load impedance for an output transistor is $90\,\Omega$. Calculate the turns ratio of the output transformer required to match the transistor to a $8\,\Omega$ load.

Solution
From equation (5.1) $n = \sqrt{\frac{90}{8}} = 3.35$ (*Ans.*)

D.C. stabilization of a transistor amplifier is best achieved by means of the potential divider bias circuit, but to minimize d.c. power losses, the emitter resistance should be of low value, perhaps as small as $1\,\Omega$. When such a low value of emitter resistance is used, the resistor is not decoupled because the required capacitance value would be very high. In amplifiers with a power output of several watts and a collector current of several amperes the emitter resistance may sometimes be omitted.

Fundamentally, a power amplifier is a converter of d.c. power taken from the collector power supply into a.c. power delivered to the load. Usually, the amplifier parameter of the greatest importance is not its power gain but the efficiency of the power conversion. The collector efficiency η is defined as

$$\eta = \frac{\text{a.c. power output to load}}{\text{d.c. power taken from power supply}} \times 100\% \qquad (5.2)$$

The d.c. power P_{dc} taken from the power supply is equal to the product of the collector supply voltage and the direct component of the collector current, i.e.

$$P_{dc} = E_{cc}I_{c(dc)} \qquad (5.3)$$

The d.c. power input to an amplifier provides the a.c. power output plus various power losses within the amplifier itself. Therefore

$$P_{dc} = P_{OUT(ac)} + \text{d.c. power losses} \qquad (5.4)$$

D.C. power is lost within the amplifier because of the resistance of the primary winding of the output transformer, the emitter resistance, and dissipation at the collector of the transistor. Often the total power lost in the transformer and in the emitter resistance is small in comparison with the collector dissipation and is neglected. Neglecting these losses is the same thing as assuming that the d.c. resistances of the primary winding and the emitter circuit are zero. If this assumption is made then equation (5.4) can be written as

$$P_{dc} = O_{OUT(ac)} + P_d \qquad (5.5)$$

where P_d is the collector dissipation. Rearranging,

$$P_d = P_{dc} - P_{OUT(ac)} \qquad (5.6)$$

The d.c. power P_{dc} taken from the supply is constant and hence equation (5.6) shows that the collector dissipation attains its maximum value when the input signal is zero and there is no output power. Care must be taken in the design of a power amplifier to ensure that this maximum value of collector dissipation does not exceed the maximum safe value for the transistor—a figure quoted by the manufacturer—or the device may be damaged. Very often the transistor is mounted on a heat sink to remove heat as efficiently as possible and limit the rise in junction temperature to a safe value.

EXAMPLE 5.2

A single-ended transistor power amplifier takes a mean collector current of 1 A from a −12 V supply and delivers an a.c. power of 2.4 W to a transformer coupled load. Calculate (a) the collector efficiency, and (b) the collector dissipation if all other losses may be neglected.

Solution
From equation (5.3), $P_{dc} = 12 \times 1 = 12$ W

Therefore, from equation (5.2), $\eta = \dfrac{2.4}{12} \times 100 = 20\%$ (*Ans.*)

From equation (5.6), $P_d = 12 - 2.4 = 9.6$ W (*Ans.*)

Graphical Determination of Power Output and Efficiency

The power output and efficiency of an audio-frequency power amplifier can be determined with the aid of an a.c. load line drawn on the output characteristics of the transistor using the procedure established in the previous chapter. If the d.c. resistances of the output transformer primary winding and the emitter circuit are negligible, the d.c. load on the transistor may be taken as zero. The d.c. load line is then drawn *vertically* upwards from the point $V_{ce} = E_{cc}$ volts, as in Fig. 5.2. The operating point must lie on this d.c. load line. The a.c. load line *must* pass through the chosen operating point with a slope equal to $1/n^2 R_L$, where n is the turns ratio of the output transformer and R_L is the load impedance connected across the secondary terminals of the output transformer.

If an alternating signal is applied to the input terminals of the amplifier the base current of the transistor will vary about its steady (quiescent) value. The corresponding values of output current and voltage can then be obtained by projecting from the intersection of the a.c. load line and the appropriate base current curves to the output current and voltage axes. In Fig. 5.2 the output current is varied between a maximum value I_{max} and a minimum value I_{min} the corresponding values of

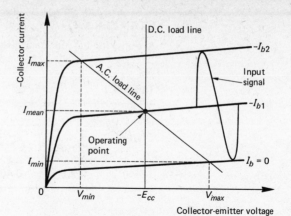

Fig. 5.2 A.C. and d.c. load lines

output voltage being V_{min} and V_{max}. The peak-to-peak swing of the output current is $I_{max} - I_{min}$, the output voltage swing is $V_{max} - V_{min}$, and so the peak output current and voltage are $(I_{max} - I_{min})/2$ and $(V_{max} - V_{min})/2$ respectively.

The a.c. power output developed in the load (assuming the transformer is 100% efficient) is equal to the product of the r.m.s. values of the a.c. components of the output current and output voltage, i.e.

$$P_{OUT} = \frac{I_{max} - I_{min}}{2\sqrt{2}} \times \frac{V_{max} - V_{min}}{2\sqrt{2}}$$

$$= \frac{(I_{max} - I_{min})(V_{max} - V_{min})}{8} \text{ watts} \qquad (5.7)$$

Alternatively, since $V_{OUT} = I_{OUT} n^2 R_L$,

$$P_{OUT} = \frac{(I_{max} - I_{min})^2}{8} n^2 R_L \text{ watts} \qquad (5.8)$$

EXAMPLE 5.3

A transistor used in a Class A audio-frequency power amplifier takes a collector bias current of 5 mA from a −10 V supply. When a sinusoidal input signal is applied to the amplifier the collector voltage varies between −2 V and −18 V and the collector current between −8 mA and −2 mA. Calculate (a) the d.c. power taken from the supply, (b) the a.c. power output, and (c) the collector efficiency.

Solution
From equation (5.3),

$$P_{dc} = 10 \times 5 \times 10^{-3} = 50 \text{ mW} \qquad (Ans. (a))$$

From equation (5.7),

$$P_{OUT} = \frac{(8-2)(18-2) \times 10^{-3}}{8} = 12 \text{ mW} \qquad (Ans. (b))$$

From equation (5.2),

$$\eta = 12/50 \times 100 = 24\% \qquad (Ans. (c))$$

EXAMPLE 5.4

A transistor has the data given in Table 5.1.

Table 5.1

V_{ce} (V)	I_b = −2 mA	−6 mA	I_c (A) −10 mA	−14 mA	−18 mA
−1	−0.02	−0.22	−0.40	−0.60	−0.80
−40	−0.20	−0.40	−0.60	−0.80	−1.00

Plot the output characteristics. The transistor is used in a single-ended transformer-coupled Class A amplifier with a load resistance of 5 Ω. The output transformer has a step-down turns ratio of 3.464:1 and negligible d.c. resistance. The collector supply voltage is −20 V, the base bias current is −10 mA, and there is no emitter resistor. Find, by drawing d.c. and a.c. load lines, the collector dissipation, the a.c. power output and the collector efficiency when a sinusoidal signal of peak value 8 mA is applied to the transistor.

Solution

The output characteristics of the transistor are shown plotted in Fig. 5.3. Since the d.c. load resistance is zero the d.c. load line has been drawn vertically upwards from the point $V_{ce} = E_{cc} = -20$ V, and the operating point, P, has been located at the intersection of the d.c. load line, the curve for a base bias current of −10 mA. The a.c. load on the transistor is $(3.464)^2 \times 5$, or 60 Ω, and so an a.c. load line of slope −1/60 must be drawn passing through the operating point. Using the method of the previous chapter, the load line for a d.c. load of 60 Ω is first drawn assuming any convenient value of supply voltage; this is the dotted load line in Fig. 5.3. The required a.c. load line is then drawn parallel to this dotted line and passing through the operating point, as shown.

By projecting from the a.c. load line it can be seen that an 8 mA peak base current produces a collector swing of 0.194 A to −0.8 A, or 0.606 A, and a collector voltage swing of −1.5 V to −38.6 V, or 37.1 V. Therefore, from equation (5.7),

$$P_{OUT} = \frac{37.1 \times 0.606}{8} = 2.81 \text{ W} \quad (Ans.)$$

From equation (5.3),

$$P_{dc} = 20 \times 0.5 = 10 \text{ W}$$

From equation (5.6)

$$P_d = 10 - 2.81 = 7.19 \text{ W} \quad (Ans.)$$

and from equation (5.2),

$$\eta = \frac{2.81}{10} \times 100 = 28.1\% \quad (Ans.)$$

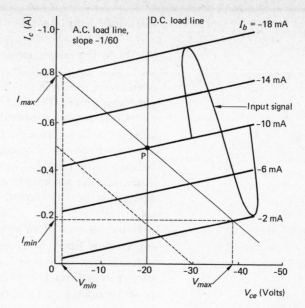

Fig. 5.3

Maximum Collector Efficiency

Expressions have previously been obtained for the a.c. power output, the d.c. power input and the efficiency of an audio-frequency power amplifier. Combining these results gives the following expression for the efficiency of power conversion using a transistor:

$$\eta = \frac{(I_{max} - I_{min})(V_{max} - V_{min})}{8E_{cc}I_{c(dc)}} \times 100\%$$

Suppose the peak value of the alternating component of the output current is equal to the direct current taken from the supply and the peak alternating output voltage is equal to the supply voltage (assuming the d.c. load on the transistor is zero). Then

$$I_{max} = 2I_{c(dc)} \quad I_{min} = 0 \quad V_{max} = 2E_{cc} \quad V_{min} = 0$$

and the collector efficiency becomes

$$\eta = \frac{2I_{c(dc)} \times 2E_{cc}}{E_{cc} \times 8I_{c(dc)}} \times 100 = 50\%$$

In practice, the collector voltage cannot be driven to zero without excessive distortion of the output waveform, and hence efficiencies of less than 50% are obtained. Transistor circuits may have efficiencies of 40% or more. In addition, the output transformer is not loss free and may have an efficiency of 80% or less.

Push-Pull Amplifiers

(1) When the output power available from a given transistor used in a single-ended circuit is inadequate the *push-pull* connection, shown in Fig. 5.4, is often used. The input and output transformers, TR_1 and TR_2, are accurately centre-trapped, and **Class A** bias is provided using the methods described in Chapter 3. Separate emitter resistors have been shown in Fig. 5.4 and have the advantage of permitting an accurate d.c. balance to be obtained between the two halves of the circuit. Alternatively, a common emitter resistor may be used; if this resistor is not decoupled the a.c. balance of the circuit is improved.

In the absence of an input signal a steady (quiescent) collector current flows in each half of the circuit, and the two currents flow in opposite directions in the two halves of the output transformer primary winding (Fig. 5.5a) tending to produce equal m.m.f.s of opposite polarity, so that d.c. saturation of the core is avoided. These m.m.f.s are completely cancelled if the two halves of the circuit are balanced. Saturation of the core would cause waveform distortion. A physically smaller core, perhaps without an air gap, may be employed, and this means that the output transformer can be both smaller and lighter than the output transformer required for a single-ended stage producing the same power output.

When a signal is applied to the input terminals of a push-pull amplifier the two transistors are driven in antiphase. Referring to Fig. 5.5b, during the half-cycle of the input signal which makes point A positive with respect to point B, the e.m.f. induced in the secondary winding of the input transformer TR_1 makes point C negative relative to point D. Point C is then negative and point D positive with respect to earth. The a.c. components of the two collector currents are then in antiphase with each another. The two signal currents flow in the same direction in the output transformer primary winding and hence the a.c. flux set up in the core is proportional to the *sum* of the two currents. The a.c. flux in the core cuts the turns of the secondary winding where it induces an alternating e.m.f., so that, if the secondary winding is closed in a load impedance, a load current flows.

Analysis of the push-pull circuit shows that all second and higher-order even-harmonic components generated by the transistors are reduced to a low level. This means that an output power of more than twice that available from one transistor can be obtained for the same distortion level. Alternatively, the same power output with a smaller percentage distortion can be obtained.

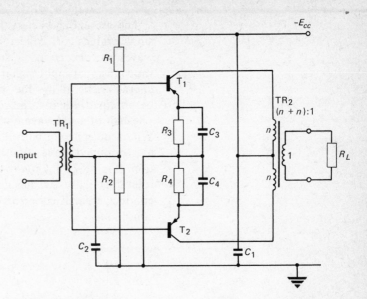

Fig. 5.4 Transformer coupled push-pull amplifier

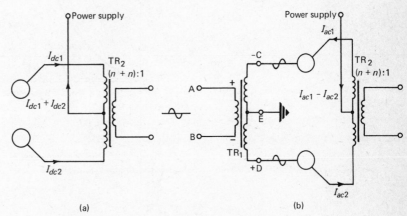

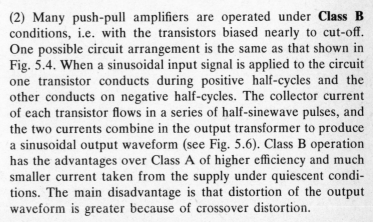

(a)   (b)

Fig. 5.5 Showing (a) the direct currents and (b) the alternating currents flowing in a push–pull circuit

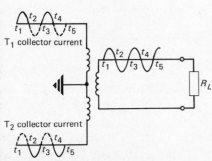

Fig. 5.6 Alternating currents in a Class B push-pull amplifier

(2) Many push-pull amplifiers are operated under **Class B** conditions, i.e. with the transistors biased nearly to cut-off. One possible circuit arrangement is the same as that shown in Fig. 5.4. When a sinusoidal input signal is applied to the circuit one transistor conducts during positive half-cycles and the other conducts on negative half-cycles. The collector current of each transistor flows in a series of half-sinewave pulses, and the two currents combine in the output transformer to produce a sinusoidal output waveform (see Fig. 5.6). Class B operation has the advantages over Class A of higher efficiency and much smaller current taken from the supply under quiescent conditions. The main disadvantage is that distortion of the output waveform is greater because of crossover distortion.

The mutual characteristics of a transistor are non-linear for small values of collector current and this gives rise to waveform distortion, as shown by Fig. 5.7a. Projecting from the sinusoidal input signal voltage waveform onto the mutual characteristic allows the waveform of the collector current to be obtained. Ideally, the collector current waveform should be one-half of a sinewave but, as can be seen, this is not the case. When the collector current waveforms of the output transistors are combined the resultant waveform exhibits crossover distortion (Fig. 5.7b). Crossover distortion can be reduced to a tolerable figure by biasing both of the output transistors to conduct a small collector current under quiescent (no-signal) conditions.

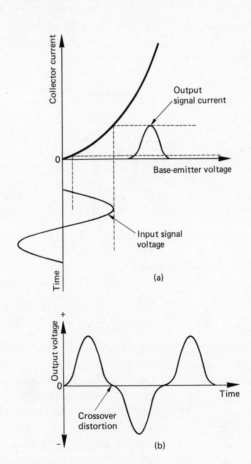

Fig. 5.7 Crossover distortion

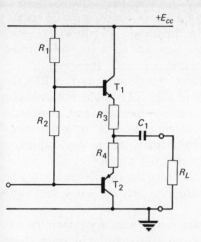

Fig. 5.8 Basic complementary pair Class B push-pull amplifier

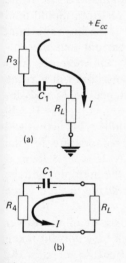

(a)

(b)

Fig. 5.9

(3) The transformer-coupled Class B circuit is nowadays generally only employed when a high output power is required. For most lower power applications, for example the output stage of domestic radio receivers, a transformer-less circuit is used. A push-pull amplifier can be built without the use of a transformer because of the availability of matched pairs of n-p-n/p-n-p power transistors. Such a circuit, known as the **complementary symmetry Class B** circuit, is shown, in basic form, in Fig. 5.8. The values of resistors R_1, R_2, R_3 and R_4 are chosen so that with zero input signal the base of transistor T_1 is held at a positive potential relative to its emitter, while the base potential of T_2 is negative with respect to its emitter. Both transistors are therefore forward biased and conduct a small collector current to minimize crossover distortion. The quiescent condition of the circuit is then with the junction of the emitter resistors R_3 and R_4 at a potential equal to one-half the collector supply voltage, i.e. $E_{cc}/2$ volts.

When an input signal is applied to the circuit its positive half cycles drive n-p-n transistor T_1 into conduction and turn p-n-p transistor T_2 off. If the amplitude of the input signal is large enough the conducting transistor is driven into saturation at the peak of the half cycle. Then the (ON) resistance of T_1 is very low and the (OFF) resistance of T_2 is very high and the circuit can be redrawn as shown in Fig. 5.9a. Current flows via R_3 and R_L to charge capacitor C_1 to a voltage of E_{cc} volts, the voltage across the load R_L is then also E_{cc} volts. During negative half cycles of the input signal voltage T_1 is turned off and T_2 is caused to conduct. At the peak of the half cycle T_2 is driven into saturation and then the circuit can be represented by Fig. 5.9b. Capacitor C_1 now discharges through resistors R_4 and R_L in series, the current flowing in R_L in the opposite direction to before. When C_1 has completely discharged, the voltage across R_L is zero. The load voltage varies about its mean value of $E_{cc}/2$ volts, reaching a maximum positive value of E_{cc} volts and a minimum value of zero. The peak value of the a.c. component of the load voltage is $E_{cc}/2$ volts and the maximum output power of the circuit is

$$\left(\frac{E_{cc}}{2\sqrt{2}}\right)^2 \Big/ R_L \quad \text{or} \quad E_{cc}^2/8R_L \text{ watts}$$

If the amplitude of the input signal voltage is not large enough to drive the transistors into saturation the output power will be less than $E_{cc}^2/8R_L$ watts.

The output transistors could be connected with their collectors commoned instead of their emitters and the circuit would operate in a similar manner. There are a number of disadvantages associated with this possible alternative connection, however, and these mean it is rarely, if ever, used. These disadvantages are as follows: (i) when the output transistors are connected with common emitter they act as a pair of emitter followers and the 100% negative feedback then applied reduces distortion of the output waveform; (ii) the BOOT-STRAP arrangement shown in Fig. 5.11 could not be used with commoned collectors because of phase and amplitude difficulties; (iii) the low output impedance of an emitter follower makes matching the output transistors to the load much easier with common emitters; and (iv) a practical circuit usually includes negative feedback from the output transistors to the driver stage, which is harder to arrange when the common collector circuit is used.

The correct operation of the complementary symmetry circuit depends upon the quiescent potential at the junction of R_3 and R_4 being held constant at $E_{cc}/2$ volts. This is commonly achieved by the use of d.c. negative feedback from the junction of the emitter resistors to the base of the driver transistor, as shown in Fig. 5.10. The base bias voltage for the driver transistor T_1 is obtained from a potential divider $R_5 + R_6$ connected between the output stage midpoint and earth. If the d.c. voltage at the junction of the emitter resistors should tend to rise the base bias voltage of T_1 will increase in proportion. Then T_1 will conduct a larger collector current and the voltages dropped across resistors R_1 and R_2 will increase. As a result the base potentials of the output transistors will become less positive. This will increase the resistance of T_2 and decrease the resistance of T_3. The voltage dropped across T_2 will therefore rise while the voltage across T_3 falls. This action will reduce the midpoint voltage tending to compensate for the original increase. The action of the d.c. feedback loop is equally effective in counteracting a fall in the midpoint voltage and so its effect is to stabilize the voltage at the desired value of $E_{cc}/2$ volts.

The d.c. component of the collector current of the driver transistor T_1 must be greater than the peak base current taken by the output transistors T_2 and T_3. Because of this the maximum possible values of resistors R_1 and R_2 are restricted and this, in turn, sets a top figure to the gain of the driver stage. An increase in gain can be achieved if the stage is *bootstrapped*, as shown in Fig. 5.11. The bias resistor R_1 has been split into two parts R_{1a} and R_{1b} and their junction has been connected to the top of the load resistor R_L. When a signal is applied the emitter potentials of the output transistors T_2 and T_3 vary in sympathy and so does the junction of R_{1a} and R_{1b}. This means that signal voltages that are very nearly equal—because the voltage gain of an emitter follower is approximately unity—exist at each end of R_{1b}. The signal

$R_1 > R_2$

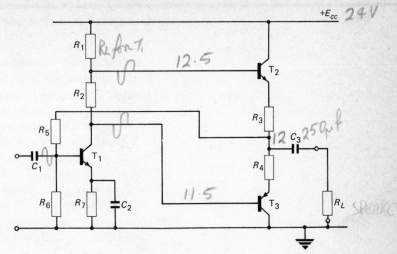

$+E_{cc}$ 24V

R_1 R₂ for T₁

12.5

T_2

R_2

R_5

R_3

12 C_3 250μf

C_1

T_1

R_4

11.5

R_6 R_7 C_2

T_3

R_L SPEAKER

Fig. 5.10 Complementary symmetry Class B push-pull amplifier

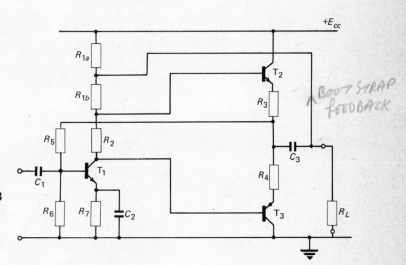

$+E_{cc}$

R_{1a}

T_2

↑ BOOTSTRAP FEEDBACK

R_{1b}

R_3

R_5 R_2

C_1

T_1

C_3

R_4

R_6 R_7 C_2

T_3

R_L

Fig. 5.11 The bootstrapped Class B amplifier

frequency current that flows in R_{1b} is therefore very small and so the effective a.c. resistance of the component is very high. Since the a.c. voltage gain of a transistor depends upon its a.c. collector load impedance the driver stage is able to have a greater gain than would otherwise be possible.

A number of varieties of the Class B complementary symmetry amplifier just described are in common use. The output stage emitter resistors may be omitted although there is then a greater likelihood of the output transistors suffering damage if, for some reason, they attempt to pass an excessive current. The methods shown of obtaining stabilization of d.c. conditions of the output stage and bootstrapping may also be varied.

Integrated Power Amplifiers

The maximum output power that can be obtained from a monolithic audio power amplifier is determined by the maximum internal power dissipation which the device is able to handle. To increase its power handling capability an integrated circuit may be mounted on a suitable heat sink to ensure the rapid and efficient removal of heat.

Typically, one integrated power amplifier can develop 2.5 W output power without an external heat sink, and up to 5 W when an external heat sink is supplied, across a 16 Ω load. The total percentage harmonic distortion is about 10% and the 3 dB bandwidth is 60 Hz to 20 kHz. If the device is operated to produce a smaller output power the distortion figure is reduced and can be lowered to about 0.5%. The circuit has an efficiency of 75% and a gain of 40 dB. Another integrated power amplifier which is currently available offers the following data: supply voltage range 10–20 V, output power in a 8 Ω load 3 W with a sensitivity of 10 mV. The total harmonic distortion is then 10% but if the output power is restricted to 1 W the total distortion falls to 1%.

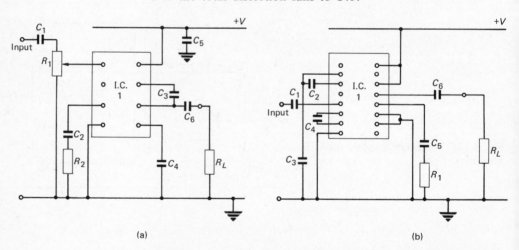

Fig. 5.12 Two integrated audio frequency power amplifiers

Some typical circuits using integrated audio power amplifiers are shown in Fig. 5.12. For both amplifiers a number of external components must be fitted to the appropriate terminals of the integrated circuits. The function of the components shown in Fig. 5.12a are as follows: C_1 and C_6 are coupling, and C_4 and C_5 are decoupling capacitors; C_2 and R_2 in series are frequency compensation components provided to ensure circuit stability; C_3 provides bootstrapping of the output signal; and, finally, the variable resistor R_1 acts as a volume control. The externally fitted components in Fig. 5.12b perform the following functions: C_1 and C_6 are the input and output coupling capacitors; C_2 couples two parts of the internal i.c. circuitry together; C_3 is a high-frequency decoupling component; C_4 adjusts the gain/frequency characteristic of the amplifier; and R_1/C_5 provide stability.

Discrete component technology must still be used, however, when power outputs beyond the capability of present i.c. power amplifiers are required. Further, a lower percentage distortion figure for a given power output is still obtainable from a correctly designed and adjusted discrete component circuit, and so any high-performance design may, perforce, rule out the use of an a.f. power amplifier i.c.

Measurements on Power Amplifiers

The main parameters of a power amplifier that can be measured are: (i) the maximum power output, (ii) the sensitivity, (iii) the power gain, (iv) the gain/frequency characteristic, and (v) distortion of the output waveform. The methods employed to measure the first four parameters are essentially the same as were discussed for a small-signal amplifier in Chapter 3. Provided the input and load impedances of the amplifier are known, the corresponding values of power and power gain can be readily calculated. Alternatively, the output power can be measured directly by means of an output power meter; this is an instrument which can be switched to have the same impedance as the amplifier load which it replaces when a measurement is made.

Distortion of the output signal waveform is the result of non-linearity in the transfer characteristic of the amplifier. The two major constituents of non-linearity distortion are *harmonic* and *intermodulation* distortion and were discussed in Chapter 4. Here the intention is to consider the ways in which these two kinds of distortion can be measured.

Measurement of Harmonic Distortion

There are two ways in which the percentage harmonic distortion of a waveform can be measured. With one method the total amplitude of the harmonic components is measured directly using a *distortion factor meter*. In the second method the amplitude of each harmonic component is separately measured by a *harmonic analyser.*

The DISTORTION FACTOR METER allows a rapid and simple direct measurement of the percentage total harmonic distortion of a waveform to be made. The waveform to be measured is applied to the input terminals of the instrument and its r.m.s. voltage is measured. A filter is then switched into a circuit that removes the fundamental frequency component of the waveform. The r.m.s. voltage of the distortion components is then measured. The ratio of the two readings is then equal to the ratio

$$\frac{\text{r.m.s. distortion voltage}}{\text{r.m.s. distortion plus signal voltage}}$$

and this, when multiplied by 100, is taken as being the percentage harmonic distortion. This is not strictly true, since the percentage distortion is actually

$$\frac{\text{r.m.s. distortion voltage}}{\text{r.m.s. signal voltage}} \times 100\%$$

but the error involved is small.

A HARMONIC ANALYSER operates on a very similar principle to a superheterodyne radio receiver that can be tuned over a frequency band of 0 Hz to, say, 50 kHz. The analyser has a very narrow band i.f. amplifier (3 dB bandwidth about 10 Hz) whose output voltage is detected, and the d.c. output is applied to an output meter. The waveform to be measured is applied to the input terminals of the analyser and the instrument is tuned to the fundamental frequency. The gain is then adjusted until a convenient indication is obtained on the meter; generally this is either full-scale deflection or 0 dB. The analyser is then tuned to each harmonic frequency in turn and each time the reading of the output meter is noted. In this way the amplitude of each harmonic component relative to the fundamental amplitude is determined. The percentage total harmonic distortion is then given by

$$[V_2^2 + V_3^2 + V_4^2 + \ldots] \times 100\%$$

where V_2, V_3 etc, are the relative amplitudes of each component. Alternatively, a spectrum analyser can be used [RSIII].

Measurement of Intermodulation Distortion

Measurement of intermodulation distortion in an amplifier is carried out by applying a two-frequency signal to the input of the amplifier and detecting the additional component frequencies which then appear at the output. The methods of carrying out the test have become standardized, and are known, respectively, as the C.C.I.T.T. and the S.M.P.T.E. methods.

Fig. 5.13 Measurement of intermodulation distortion (a) C.C.I.T.T. method, (b) S.M.P.T.E. method

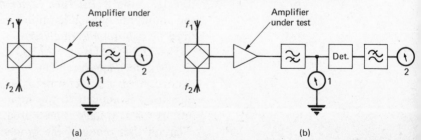

(a)   (b)

The test circuit used to measure intermodulation using the C.C.I.T.T. method is shown in Fig. 5.13a. Two sinusoidal voltages, of equal amplitude and frequencies f_1 and f_2 (with $f_1 = 1.1f_2$) are combined in a hybrid coil and applied to the input terminals of the amplifier. The voltage output of the amplifier is measured by the peak-responding voltmeter 1. The output signal is then applied to the low-pass filter which removes the input frequencies f_1 and f_2. Thus, only the modulation products f_2+f_1, $2(f_2 \pm f_1)$, $3(f_2 \pm f_1)$ etc. pass through the filter and their resultant is measured by the peak-responding voltmeter 2. The percentage intermodulation distortion is then

$$100 \times \frac{\text{meter 2 reading}}{\text{meter 1 reading}}$$

Fig. 5.13b shows the equipment used for the S.M.P.T.E. method of intermodulation measurement. Two sinusoidal input signals are again employed but now $f_2 = 10f_1$ and the amplitude of the low frequency voltage is four times greater than the amplitude of the higher frequency signal. The output of the amplifier is connected to a high-pass filter which removes frequency f_1. The output of the filter is measured by peak-responding meter 1 and is then detected. The detected output is passed through a low-pass filter and then measured by meter 2. The percentage intermodulation distortion is now given by

$$100 \times \frac{\text{indication of meter 2}}{\text{indication of meter 1}}$$

Exercises

5.1. (a) List the function of each component shown in the Class B push-pull amplifier of Fig. 5.14. (b) List the advantages and disadvantages of Class B operation compared with Class A.

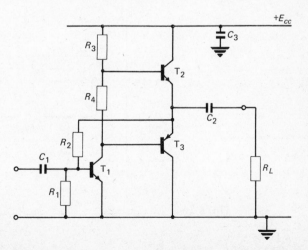

Fig. 5.14

5.2. (a) List the function of each component shown in the integrated power amplifier circuit shown in Fig. 5.15.

(b) List the advantages of using integrated circuits instead of circuits made up of discrete components.

(c) Quote typical figures for power output, sensitivity and 3 dB bandwidth for an amplifier of this type.

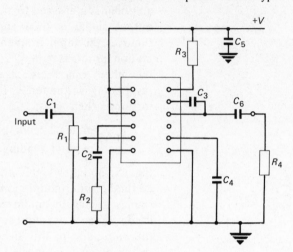

Fig. 5.15

5.3. Fig. 5.16 shows the circuit diagram of an audio-frequency power amplifier.

(a) What are the functions of R_9, C_3, C_2 and R_3?

(b) Which components provide bootstrapping?

(c) Is negative feedback applied to the circuit?

(d) What is a typical efficiency for such a circuit?

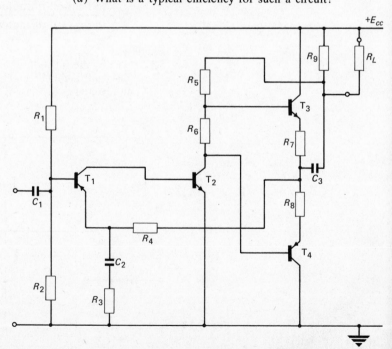

Fig. 5.16

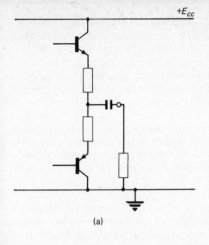

(a)

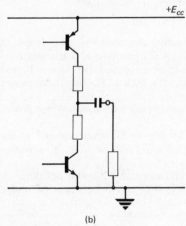

(b)

Fig. 5.17

5.4. Sketch a circuit diagram of a Class B push-pull amplifier which uses a complementary pair of transistors. Describe the operation of the complete circuit. Show by the use of appropriate characteristic curves how crossover distortion occurs and explain how it may be reduced. What type of distortion is introduced if the characteristics of the devices used are not identical?

5.5. (a) Figs. 5.17a and b show two possible methods of connecting a complementary pair of transistors to form an elementary push-pull amplifier. Give three reasons why the circuit in Fig. 5.17a is superior to the circuit in Fig. 5.17b. (b) For the circuit in Fig. 5.17a, describe the complete load current path for each half cycle of output and draw graphs of input voltage, output voltage and supply current to a common time scale. (c) If the load has a resistance of 8 Ω, determine a suitable value for capacitor C when the circuit is used as an audio power amplifier. (C&G)

5.6. (a) A class A power amplifier is shown in Fig. 5.1. The collector current alternates between 3 mA and 110 mA and its quiescent value is 58 mA. $E_{cc} = -12$ V. The load resistance of 12 Ω when referred to the primary of the transformer is 325 Ω. Estimate, stating assumptions made, (i) the transformer ratio n, assuming an ideal transformer, (ii) the a.c. power output, (iii) the minimum collector rating for the transistor, (iv) the d.c. power supplied, neglecting the base current, (v) the percentage efficiency.

(b) Explain why a greater efficiency is achieved by transformer coupled resistive loading rather than direct resistive loading. (C&G)

5.7. (a) With the aid of a circuit diagram, describe the operation of an audio-frequency push-pull amplifier. (b) List three advantages of push-pull operation over the use of a single amplifying device. (c) Explain where a push-pull stage would most effectively be used in a multi-stage amplifier. (C&G)

5.8. (a) Sketch a set of typical collector characteristics for a power transistor. For a power transistor supply voltage marked as V_{cc} on the characteristics show typical load lines for (i) a load resistor (ii) a transformer-coupled resistive load. (b) A power transistor is supplied from a 15 V power supply and feeds a 10 Ω resistance through a 2:1 step-down transformer. The collector current varies sinusoidally between 650 mA and 50 mA. Determine (i) the output power, (ii) the power taken from the supply, (iii) the efficiency, and (iv) the collector dissipation. (C&G)

5.9. (a) Which electrode of a power transistor is usually connected to the case? Explain why this is so and discuss the complications which this can cause in the circuitry or in the transistor mounting. (b) Why are heat sinks often used with power transistors? (c) The transistor in a Class A power amplifier is mounted so that its maximum permitted collector dissipation is 15 W. With the signal applied the stage has an efficiency of 35%. Estimate the maximum output power if the signal is continuous. (C&G)

```
 o1        16o
 o2        15o
 o3        14o
 o4        13o
 o5        12o
 o6        11o
 o7        10o
 o8         9o
```

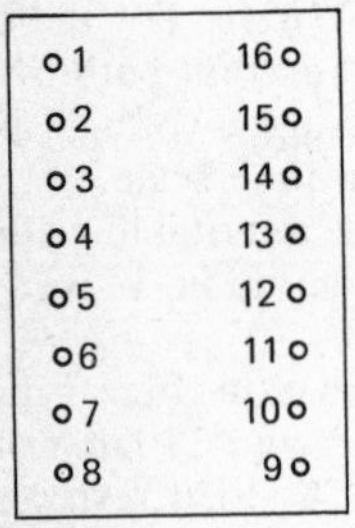

Fig. 5.18

1	+V
2–3	not used
4–5	earth
6	bootstrap
7	frequency compensation
8	negative feedback
9	not used
10	input
11–12	not used
13–14	earth
15	not used
16	output

5.10. (*a*) Explain why the internal dissipation of a Class A amplifier rises when the input signal is reduced in amplitude. (*b*) A Class A power amplifier has a 4 Ω transformer-coupled load. With the maximum load power of 2 W the efficiency is 35%. Assuming the output signal to be sinusoidal, (i) calculate the internal dissipation of the amplifier when the output power is at its maximum, (ii) plot a curve showing the variation of internal dissipation with load voltage as the input signal is reduced to zero. (C&G)

5.11. Explain how you would measure the total percentage distortion at the output of an audio-frequency power amplifier using (*a*) a distortion factor meter and (*b*) a harmonic analyser. When a particular waveform was measured using each instrument in turn the distortion factor meter gave the higher reading. Explain the reasons for this.

5.12. Fig. 5.18 shows the function of the terminals of an integrated circuit. Draw a circuit diagram of an audio-frequency power amplifier using this device. List the advantages of your circuit over one using discrete components throughout.

Short Exercises

5.13. Explain, with the aid of a mutual characteristic, what is meant by Class A, Class B, and Class C operation of a transistor.

5.14. What is meant by the terms (i) single-ended, (ii) push-pull, and (iii) complementary in connection with audio-frequency power amplifiers?

5.15. List typical performance characteristics of a monolithic power amplifier.

5.16. What is meant by the following terms when applied to an audio-frequency power amplifier (i) optimum load, (ii) sensitivity, (iii) heat sink?

5.17. A power amplifier has an efficiency of 22% and develops an output power of 2.5 W. (i) Is it likely to be operated under Class A or Class B conditions? (ii) Calculate the d.c. power taken from the collector power supply.

5.18. An amplifier has an efficiency of 48% and a collector dissipation of 0.92 W. Calculate the a.c. output power of the amplifier.

5.19. Explain what is meant by crossover distortion.

5.20. A Class B complementary symmetry push-pull amplifier operates from a 30 V power supply. Calculate the maximum a.c. power it can deliver to an 8 Ω load.

6 Tuned Amplifiers

Principles of Operation

A tuned amplifier is one which is required to handle a relatively narrow band of frequencies centred on a particular radio frequency. Such an amplifier has two main functions: it must provide a specified gain over a given frequency band and it must provide the selectivity necessary to ensure that frequencies outside the wanted band are not amplified to the same extent. The required selectivity is generally obtained by a parallel-resonant circuit acting as the collector or drain load for the transistor or f.e.t. The required LC product, for resonance to occur at the desired centre frequency, is fairly small and is easy to obtain; the stray and transistor capacitances that adversely affect the high-frequency performance of an untuned amplifier now contribute to the necessary tuned circuit capacitance. High-power radio-frequency tuned amplifiers must employ thermionic valves as the active device since the power dissipated within the device may well be considerably greater than the greatest power any present-day transistor is capable of handling. Tetrode valves are those normally used unless the output power is very large when a triode, connected in the earthed-grid configuration, is used.

Transistors are subject to internal feedback of energy at radio frequencies and a transistor tuned amplifier must be designed to avoid instability. Generally, the common-emitter configuration is chosen in preference to the common base because it provides the greatest gain, and its input and output impedances have more convenient values. The common-base connection is sometimes chosen when one or more of its particular characteristics are required. These are as follows: (a) it gives a more or less constant gain over a wide bandwidth, (b) the spread in the current gains of different transistors of the same type is smaller since $h_{fb} = h_{fe}/(1 + h_{fe})$; and (c) a transistor in common base can provide a higher gain at frequencies near the f_t of the transistor than the same transistor connected in common emitter [E II]. Often, for example, common-base tuned circuits are employed in the input (radio-frequency) stage of v.h.f. radio receivers.

Field effect transistors can also be used to provide radio-frequency amplification and may be operated in either the common-source or the common-gate configurations. The use of a f.e.t. offers advantages over the bipolar transistor in that it generates less noise and is less likely to produce *cross-modulation*. The common-source connection has a high input impedance and can give both good gain and good noise performance. For the utmost stability at the higher frequencies the common-gate connection is sometimes used, at the expense, however, of lower gain and slightly more noise. F.E.T. tuned amplifiers are sometimes used in the tuner sections of frequency-modulation receivers and various other kinds of v.h.f./u.h.f. equipment.

A number of integrated circuit radio-frequency amplifiers are available although in many cases the package also includes a number of other circuit functions. For example, one readily available i.c. consists of an r.f. amplifier, a mixer, an i.f. amplifier, and a detector.

One of the main problems in the design of a transistor or f.e.t. tuned amplifier is the internal feedback of the device. Because of this feedback a transistor or f.e.t. may be capable of oscillation over a fairly wide range of frequencies when connected between particular source and load impedances. Another undesirable effect of the internal feedback is that it makes the input impedance of the device a function of frequency, and this in turn, makes alignment of a multistage circuit difficult.

In the past a technique known as unilateralization was often employed to cancel out the unwanted internal feedback. Nowadays, the design of a modern r.f. transistor is such that little internal feedback takes place and, at lower radio frequencies at least, circuits can be designed for maximum power transfer between stages. Any r.f. transistor will be stable if it operates between source and load impedances which are smaller than the matched impedances needed to achieve maximum power gain. The impedance values necessary to achieve stability are obtained by the use of input and output networks of suitable impedance ratios.

Single-tuned Amplifiers

The circuit diagram of a transistor tuned amplifier is shown in Fig. 6.1. The collector tuned circuit must provide the required impedance/frequency characteristic and also the necessary impedance transformation. This type of circuit is generally required to be tunable to any frequency within a given frequency band and therefore employs the mismatch method of providing stability. The tuned circuit can have either its inductance or its capacitance branch tapped [RS II] to give the required impedance transformation; in Fig. 6.1 a capacitive tap has been used.

Fig. 6.2 shows the circuit of a common-source f.e.t. radio-frequency amplifier; the drain tuned circuit provides the necessary selectivity and may sometimes be shunted by a resistor to give the required stability.

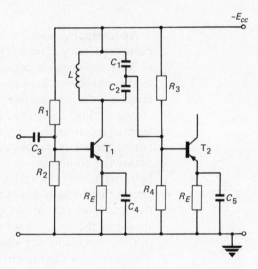

Fig. 6.1 Collector-tuned transistor amplifier

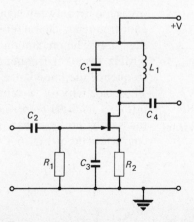

Fig. 6.2 Drain-tuned f.e.t. amplifier

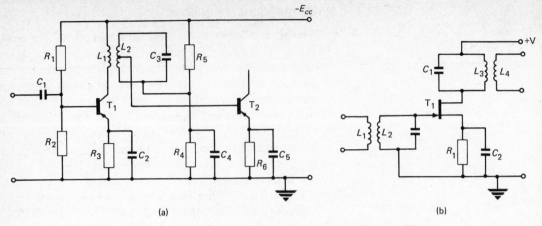

Fig. 6.3 Single-tuned transformer amplifiers

An alternative method of coupling two stages is the use of a radio-frequency transformer (Fig. 6.3). The primary winding of the transformer is connected in the collector or drain circuit of the amplifying device used and has a low d.c. resistance to minimize the d.c. voltage dropped across it. In the transistor circuit of Fig. 6.3a the secondary winding is tuned and the base terminal of the following transistor T_2 is connected to a tap on the secondary winding. T_2 cannot be directly connected across the secondary winding because its low input resistance (approximately $1000\,\Omega$) would so heavily dampen the tuned circuit that little, if any, selectivity would remain. An alternative arrangement would be to tune the primary winding as shown in the f.e.t. circuit of Fig. 6.3b.

The gain/frequency characteristic of a tuned amplifier is determined by the selectivity characteristic of its tuned circuit. The selectivity of a resonant circuit, and hence of the tuned amplifier whose collector or drain load it is, is its ability to discriminate between signals at different frequencies. A typical gain/frequency characteristic for a tuned amplifier is shown in Fig. 6.4. The maximum voltage gain of 150 occurs at the 5 MHz resonant frequency of the tuned circuit, since at this frequency the impedance of the circuit has its maximum value. The selectivity of a tuned amplifier is usually expressed in terms of its 3 *dB bandwidth*. This is the bandwidth over which the gain is not less than $1/\sqrt{2}$ times the resonant gain; for the characteristic of Fig. 6.4 the 3 dB bandwidth is 100 kHz.

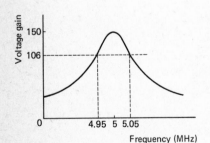

Fig. 6.4 Gain/frequency characteristic of a tuned amplifier

EXAMPLE 6.1

A tuned r.f. amplifier has the gain/frequency characteristic indicated by the data of Table 6.1.

Table 6.1

Frequency (MHz)	10.45	10.5	10.6	10.7	10.8	10.9	10.95
Gain(dB)	0	80	198	200	198	80	0

Plot the gain/frequency characteristic and determine the 3 dB, 6 dB, and 20 dB bandwidth of the amplifier.

Solution
The required characteristic is shown plotted in Fig. 6.5. From the characteristic the required bandwidths are

$$3\,dB = 324\,kHz, \qquad 6\,dB = 374\,kHz, \qquad 20\,dB = 474\,kHz$$

(*Ans.*)

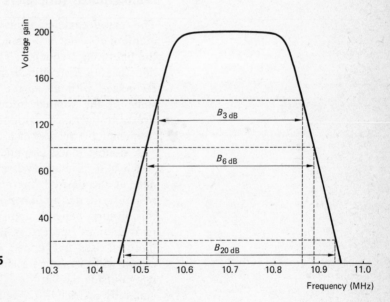

Fig. 6.5

Dual gate m.o.s.f.e.t.s are increasingly employed as r.f. amplifiers in modern equipment since the signal and bias circuits can be kept separate, permitting the optimum design of each. A typical circuit arrangement is shown in Fig. 6.6. It can be seen that the signal voltage is applied to gate terminal one while the d.c. bias voltage is at gate terminal two.

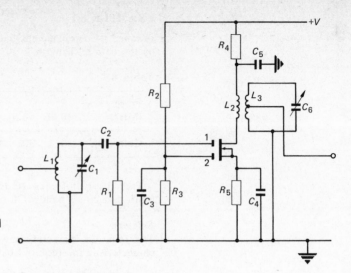

Fig. 6.6 Dual-gate m.o.s.f.e.t. tuned amplifier

Double-tuned Amplifiers

The gain/frequency characteristics of the tuned amplifiers mentioned so far are rounded and fall away on either side of the operating frequency. This means that a single-tuned amplifier cannot discriminate against unwanted frequencies near resonance without at the same time discriminating against some of the wanted frequencies. This disadvantage can be overcome in tuned amplifiers designed to work at a constant frequency by the use of a double-tuned amplifier (Fig. 6.7).

A double-tuned amplifier employs transformer coupling in which both primary and secondary circuits are tuned to resonate at the desired operating frequency. The gain/frequency response of the amplifier is that of the r.f. transformer, and this, in turn, depends upon the mutual inductance between the two windings (see Fig. 6.8). If critical coupling is employed a more or less flat-topped characteristic is obtained and the circuit will discriminate sharply against unwanted frequencies lying outside the flat top. The use of double-tuned amplifiers is generally restricted to fixed-frequency applications such as i.f amplifiers in radio receivers, because of difficulties associated with the need for the simultaneous tuning of two coupled tuned circuits.

The input resistance of a common-emitter connected transistor is of the order of 1500Ω and this figure is sufficiently low to ruin the selectivity characteristic of the coupled circuits if shunted directly across the secondary winding. It is necessary therefore, for the transistor to be connected to a tap on the secondary winding, as shown in Fig. 6.7.

BOULOGNE RADIO   FFB
NORDDEICH RADIO   DAN
KIEL RADIO   DAO
SCHEVENINGEN RADIO   PCH
ANTWERP RADIO   OSA
OOSTENDE RADIO   OSU - OST
SANDJIZA

	CITY	PORT	RIVER
1. WICK/GKR	NONE	NONE	NONE
2. STONEHAVEN/GND	EDINBURGH	LEITH	FIRTH OF FORTH
3. CULLERCOATS/GCC	NEWCASTLE	NEWCASTLE	TYNE
4. HUMBER/GKZ	HULL	HULL	HUMBER
5. NORTHFORELAND/GNF	LONDON	LONON	THAMES
6. NITON/GNI	SOUTHAMPTON	SOUTHAMPTON	SOLENT
7. LANDSEND/GLD	NONE	NONE	NONE
8. ILFRACOMBE/GIL	AVONMOUTH	AVONMOUTH	SEVERN
9. ANGLESEY/GLV	LIVERPOOL GLASGOW	LIVERPOOL	MERSEY
10. PORTPATRICK/GPK	HEYSHAM	HEYSHAM	NONE CLYDE
11. MALINHEAD/EJM	BELFAST	BELFAST	NONE
12. VALENTIA/EJK	NONE	NONE	NONE
13. PORTISHEAD/GKA	BRISTOL	NONE	AVONE

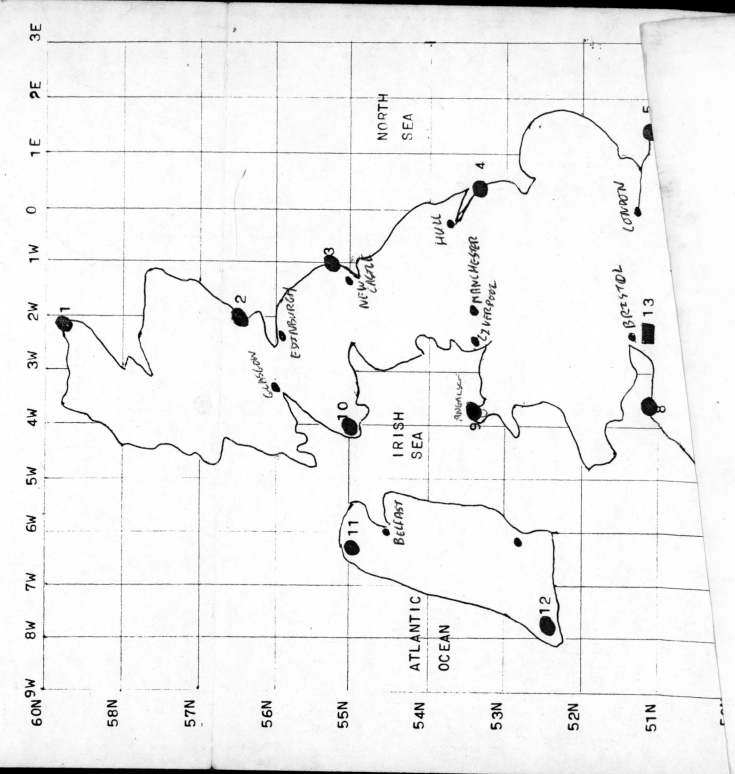

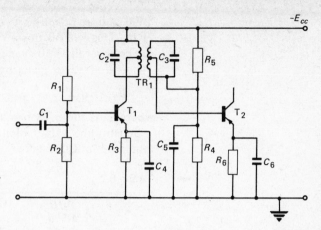

Fig. 6.7 Double-tuned transistor amplifier

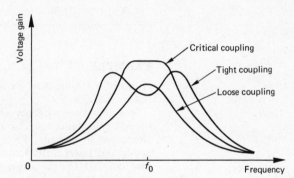

Fig. 6.8 Gain/frequency characteristic of a double-tuned amplifier

Cascaded Stages

Very often the gain required from an amplifier is greater than the gain that can be obtained from a single stage. Then, two or more stages must be cascaded to obtain the desired gain. The overall gain A_v of a multi-stage amplifier is the product of the individual stage gains; for example, if the three stages of a particular amplifier had voltage gains of 22, 27 and 20 respectively the overall voltage gain A_v would be $22 \times 27 \times 20$ or 11 880. Unfortunately, cascading stages to increase the available voltage gain also has the effect of reducing the 3 dB bandwidth of the amplifier.

Consider, for example, an amplifier having 3 dB frequencies of f_1 and f_2 respectively. For two stages the overall gain at resonance is 20^2 or 400 and the gain at f_1 and at f_2 is $(20\sqrt{2})^2$ or 200. The gain at f_1 and f_2 is only one half of the resonant gain and this means that, for the two cascaded stages, f_1 and f_2 are the 6 dB frequencies. Obviously, the 3 dB bandwidth is now less than $f_2 - f_1$, i.e. the bandwidth has *shrunk*. When three stages are considered the bandwidth shrinkage is even more pronounced. At resonance the overall gain is 20^3 or

8000 and at f_1 and at f_2 the overall gain is only $(20/\sqrt{2})^3$ or 2828. This is 9 dB down on the resonant gain.

The overall 3 dB bandwidth B_0 of an amplifier with n *identical* stages is given by

$$B_0 = B\sqrt{(2^{1/n} - 1)} \qquad (6.1)$$

where B is the 3 dB bandwidth of each stage.

Substituting values of n in equation (6.1) gives the results tabulated in Table 6.2.

Table 6.2

n	B_0/B	n	B_0/B
1	1.000	4	0.435
2	0.644	5	0.386
3	0.509	6	0.350

EXAMPLE 6.2

A tuned r.f. amplifier stage has the gain/frequency characteristic shown in Table 6.3.

Plot the gain/frequency characteristic of the stage and also the overall gain/frequency characteristic of two and three such stages connected in cascade. From the characteristics determine (*a*) the upper and lower 3 dB frequencies of one, two and three stages in cascade, and (*b*) the overall gain of two and three stages at the upper and lower 3 dB frequencies of a single stage.

Table 6.3

Fre-quency (kHz)	950	960	970	980	990	1000	1010	1020	1030	1040	1050
Gain	1.98	2.45	3.12	4.47	7.07	10	7.07	4.47	3.12	2.45	1.98

Solution
Squaring the gain figures for the single stage to obtain the overall gain of two stages, and cubing for the overall gain of three stages gives Table 6.4.

Table 6.4

Fre-quency (kHz)	950	960	970	980	990	1000	1010	1020	1030	1040	1050
Gain 1 stage	1.98	2.45	3.12	4.47	7.07	10	7.07	4.47	3.12	2.45	1.98
Gain 2 stages	3.92	5.91	9.73	19.98	50	100	50	19.98	9.73	5.91	3.92
Gain 3 stages	7.76	14.36	30.63	89.31	353.5	1000	353.5	89.31	30.36	14.36	7.76

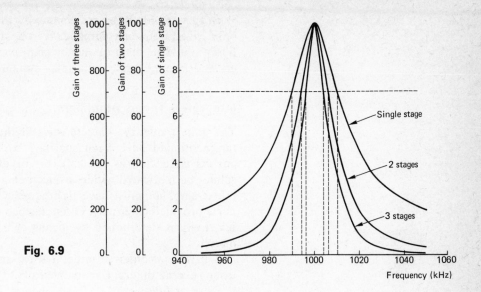

Fig. 6.9

These characteristics are shown plotted in Fig. 6.9.

(*a*) From the characteristics, the upper and lower 3 dB frequencies are

    (i) single stage: 990 and 1010 kHz
    (ii) two stages: 994 and 1006 kHz
    (iii) three stages: 996 and 1004 kHz

(*b*) At the upper and lower 3 dB frequencies of a single stage, i.e. 990 and 1010 kHz the gain of (iv) two stages is 50 and (v) of three stages is 353.5.

Stagger Tuning

When a wide bandwidth r.f. amplifier is to be constructed, the bandwidth shrinkage effect which occurs when a number of stages are *synchronously* tuned is a serious disadvantage. For a multi-stage amplifier to have a given overall bandwidth the individual stage bandwidths must be considerably wider. This means that the Q-factor of the tuned circuit of each stage must be fairly low and because of this the gain of each stage is reduced. The practical result is that the increase in overall gain obtained when an extra stage is added will be less than anticipated. Better results can be obtained if the stages are not all tuned to the same frequency. The idea of *stagger tuning* is illustrated by Fig. 6.10. Two stages are tuned to frequencies spaced equally either side of the wanted centre frequency f_0 and provided their bandwidths are suitably chosen, the overall response is reasonably flat over the operating bandwidth. When three stages are employed, one stage is tuned to the desired operating frequency with the other two stages tuned one above and one below that frequency.

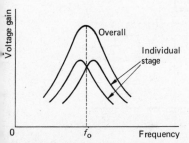

Fig. 6.10 Stagger tuning

Stagger-tuned stages are commonly used whenever a wide-band, fixed frequency amplifier is to be designed. Some examples are intermediate frequency amplifiers in television receivers and in u.h.f./s.h.f. radio-relay systems.

Integrated Tuned Amplifiers

The gain/frequency characteristics of the tuned amplifiers so far mentioned have been specified by the parallel-resonant circuits connected as the collector or drain loads. Inductors cannot be fabricated within a silicon chip and so an integrated tuned amplifier must have its frequency-determining components provided externally. Often, instead of a tuned circuit, the selectivity is determined by means of a ceramic or a crystal filter.

A variety of different integrated r.f. amplifiers are available from several different manufacturers. Parameters of typical i.c.s are as follows:

(i) Voltage gain 20 dB, bandwidth 140 MHz, maximum input signal voltage 100 mV.
(ii) Voltage gain 26 dB, bandwidth 100 MHz, maximum input signal voltage 50 mV.
(iii) Voltage gain 80 dB, maximum input signal voltage 15 μV, input impedance 15 Ω in parallel with 100 pF.

If selectivity is not provided, most i.c.s will also act as wideband (e.g. 0–3 MHz) amplifiers.

A possible i.c. r.f. amplifier circuit is shown in Fig. 6.11. The circuit uses parallel-tuned L-C circuits L_1-C_1 and L_2-C_4 to determine the centre frequency of operation. The operating frequency must, of course, lie somewhere within the bandwidth of the integrated circuit. The amplifier employs a tapped input inductor and a transformer-coupled output; C_1 and C_4

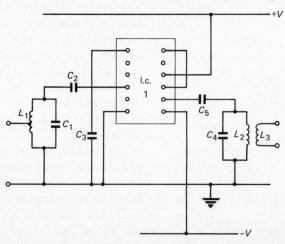

Fig. 6.11 Integrated circuit tuned amplifier

are tuning components; and C_3 decouples part of the i.c. circuitry to earth. C_2 and C_5 are d.c. blocks.

Many integrated r.f. amplifiers are associated with a number of other circuit functions within the same package. For example, one i.c. contains an r.f. amplifier, an oscillator and a mixer, while a second i.c. adds an i.f. amplifier and a detector to the functions of the first.

Tuned Power Amplifiers

Tuned radio-frequency power amplifiers find their main application in radio and television transmitters where the transmitted power may be anything from a few watts to hundreds of kilowatts. When dealing with high power levels the maximum possible efficiency is essential and for this reason tuned power amplifiers are operated under either Class C or Class B conditions. Class B operation of a valve or a transistor means that the device is biased to its cut-off point so that it conducts current only during alternate half-cycles of the input signal waveform. With Class C operation the amplifying device is biased beyond cut-off so that current flows in a series of less-than-half sine wave pulses. High-power circuits must employ tetrode or triode valves because the power dissipated internally is large but these will not be described in this book [RS II].

The basic circuits of some Class C transistor tuned power amplifiers are shown in Fig. 6.12.

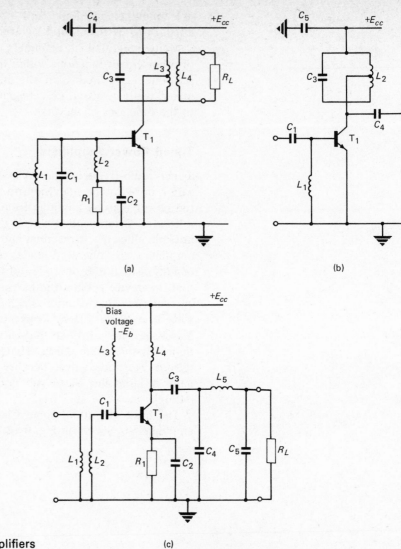

Fig. 6.12 Class C power amplifiers

In Fig. 6.12a the required Class C bias is obtained from the leaky-base circuit formed by R_1 and C_2. Inductor L_2 is a radio-frequency choke whose function is to act as a block to the flow of signal-frequency currents and prevent the signal voltage being bypassed to earth via capacitor C_2. The input circuit C_1-L_1 of the amplifier is tuned to the signal frequency and delivers the input voltage to the base terminal of transistor T_1. The positive peaks of the input signal drive the base positive and when it is more positive than 0.6 V for a silicon transistor, or 0.2 V for a germanium transistor, a collector current flows. The collector current flows for a period of time which is less than one half of the periodic time of the signal waveform (see Fig. 6.13). The collector current waveform is

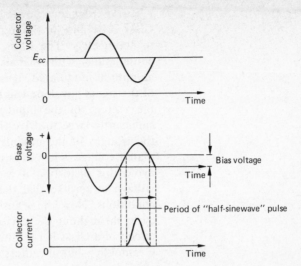

Fig. 6.13 Waveforms in a Class C amplifier

the resultant of a fundamental frequency (equal to the frequency of the input signal) plus a number of harmonics. The collector tuned circuit C_3-L_3 is tuned to the fundamental frequency. If it has a Q-factor greater than about 12 its selectivity will be sufficiently high to ensure that only the fundamental component is able to develop a voltage of any magnitude across the tuned circuit. Hence, the alternating voltage developed across the collector tuned circuit is of sinusoidal waveform (Fig. 6.13).

The voltage is coupled to the load via the mutual inductance between inductors L_3 and L_4. The inductor L_3 is tapped to ensure that the transistor works into its optimum load impedance.

The Class C circuit of Fig. 6.12b shows the coupling from the collector circuit to the load being made via a series capacitor C_4. The bias arrangement employed consists only of a radio-frequency choke L_1 connected between the base of the transistor and earth. This arrangement is only used in conjunction with very large input signal voltages; the input signal voltage is rectified by the diode formed by the base-emitter p-n junction of T_1 and the d.c. component thus produced acts as the bias voltage. Often the bias voltage can be obtained by the connection of R_1 and C_2 in the emitter circuit, as illustrated by Fig. 6.12c. When the transistor conducts, a d.c. voltage is developed across R_1 that makes the emitter of T_1 positive with respect to earth and thereby increases the reverse bias that is applied to the transistor. Fig. 6.12c also illustrates the way in which most v.h.f. and u.h.f. Class C amplifiers are coupled to their load, a π-type coupling network being used because the amplifiers normally work between either 50 Ω or 75 Ω source and load impedances.

Radio-frequency power amplifiers are generally operated in Class C because of the high efficiency obtained, in practice up to about 80%. The Class C circuit can only be employed in conjunction with signals of constant amplitude. If an amplitude-modulated signal is applied considerable distortion of the signal envelope will result. This distortion arises because the voltage of the input amplitude modulated wave is not sufficiently large to drive the transistor into conduction during the troughs of the modulation cycle, i.e. whenever the modulated voltage is less than the unmodulated carrier voltage. To avoid this distortion it is necessary to reduce the base bias voltage to zero so that the transistor operates under Class B conditions. Now the transistor will conduct once in each r.f. cycle throughout the cycle of the modulation envelope. The circuit of a Class B r.f. tuned amplifier is the same as the Class C amplifier but it suffers from the disadvantage of smaller efficiency, up to about 65%.

A modern development which allows a Class C circuit to amplify an amplitude-modulated wave is the use of overall envelope negative feedback. A fraction of the output signal envelope is fed back in antiphase with the input signal and distorts it in such a way that the distortion produced by the amplifier is cancelled out and an output with very little distortion is obtained.

Parasitic Oscillations

Sometimes a tuned power amplifier may start to oscillate at some frequency other than its normal operating frequency; oscillations of this kind are known as *parasitic oscillations*. Parasitic oscillations in a power amplifier are undesirable, even though they do not occur at the operating frequency, because they dissipate power and reduce the available wanted power output. The oscillations may also lead to the transistor being driven into operation on the non-linear part of its characteristics with a consequent increase in the distortion level at the output.

Parasitic oscillations may occur in an amplifier at a frequency which is either smaller than or greater than the frequency of amplification. Oscillations at a lower frequency generally arise because radio-frequency chokes and bypass capacitors in both the base and the collector circuits form tuned circuits which happen to be resonant at the same frequency. Then, an oscillator circuit may be formed. Consider, for example, the tuned amplifier shown in Fig. 6.14a.

The input and output tuned circuits C_1-L_1 and C_3-L_4 are each tuned to be resonant at the desired frequency of operation. L_2 and L_3 are radio-frequency chokes. C_2 is a d.c.

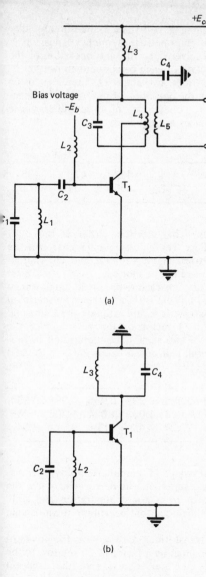

(a)

(b)

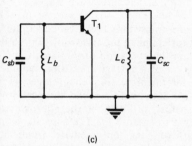

(c)

Fig. 6.14 Parasitic oscillations

blocking capacitor and C_4 decouples the collector supply. At much lower frequencies the reactances of L_1 and L_4 will be negligibly small, but now the reactances of capacitances C_2 and C_4 will not be negligible. The effective circuit of the amplifier is then as shown in Fig. 6.14b, remembering that as far as a.c. is concerned the power supplies $+E_{cc}$ and $-E_b$ are at earth potential. Providing the resonant frequencies of the tuned circuits C_2-L_2 and C_4-L_3 are equal the circuit will oscillate at that common frequency because some energy is fed from collector to base via the transistor's internal capacitance.

The obvious method of preventing parasitic oscillations is to make sure that the resonant frequencies of the parasitic tuned circuits are not equal by a suitable choice of values for the r.f. chokes and coupling/decoupling capacitors that are fitted. Parasitic oscillations may also occur at a frequency higher than the normal operating frequency of the amplifier. At higher frequencies unwanted resonant circuits may be formed by stray transistor capacitances and the inductances of connecting leads.

Consider again the circuit given in Fig. 6.14a. Stray capacitances exist in parallel with both the input and output paths while the connecting leads to the base and collector terminals possess self-inductance. At frequencies well above the operating frequency of the amplifier the tuning capacitors C_1 and C_3 and the decoupling/coupling capacitors C_2 and C_4 will all have a negligible reactance; consequently, the effective circuit of the amplifier becomes that given by Fig. 6.14c, where L_b and L_c are the self-inductances of the base and collector lead respectively, and C_{sb} and C_{sc} are respectively the stray capacitances of the base and collector circuits. This is essentially the same circuit as Fig. 6.14b and it will also oscillate if the products $C_{sb}L_b$ and $C_{sc}L_c$ are equal. To prevent parasitic oscillations L_b and/or L_c can be altered in value by shortening the length(s) of the lead(s) or perhaps C_{sb} and/or C_{sc} can be changed in value by modifying the layout of the circuit.

Exercises

6.1. Draw the circuit diagram of a two-stage transistor amplifier suitable for use in the intermediate-frequency stage of a superheterodyne radio receiver. Describe the operation of the circuit and state the function of each component shown.

6.2. Draw a circuit diagram of a two-stage i.f. amplifier using transistors. Briefly explain the operation of the biasing arrangements. Table A shows the voltage gain of an amplifier whose input and output impedances are equal.

Table A

Frequency (kHz)	450	454	462	478	498
Voltage gain	100	92	60	31	17.5

Plot the gain/frequency characteristic. What is the 3 dB bandwidth of the amplifier? The gain/frequency characteristic may be assumed symmetrical about 450 kHz and the characteristic below resonance need not be plotted. (C&G)

6.3. The voltage-gain/frequency characteristic of a single-tuned amplifying stage is given in Table B. Using your own scales on graph paper, plot this characteristic and estimate the following: (a) 3 dB bandwidth, (b) Q factor of the circuit. Plot the gain/frequency response of two such stages connected in random and indicate the overall half-power bandwidth. (C&G)

Table B

Frequency (MHz)	1	0.999	0.998	0.997	0.996	0.995	0.994
		1.001	1.002	1.003	1.004	1.005	1.006
Voltage gain	22	20.35	17.27	14.3	11.88	9.9	8.34

6.4. Sketch the circuit of a two-stage i.f. amplifier employing transistors suitable for use in a superheterodyne receiver. Give the reasons for your choice of transistor configuration and state features and components included that contribute towards gain stability.

In a receiver, each i.f. transformer stage reduces the power of a signal in an adjacent channel by a factor of 20 relative to the wanted signal. Calculate the net reduction of the adjacent channel signal if the i.f. amplifier has three such stages. (C&G)

6.5. Draw the circuit diagram of a transistor Class C tuned amplifier that employs π-networks for coupling the transistor to both its source and its load. Explain the operation of the circuit.

6.6. (a) Draw the circuit diagram of a 2-stage transistor amplifier suitable for use in the intermediate frequency stages of a communication receiver. (b) State, with reasons, the preferred configuration of the transistors. (c) What is the approximate output resistance of the first transistor? (d) The capacitance in a resonant circuit is 100 pF, the intermediate frequency 465 kHz, and the bandwidth at the -3 dB bandwidth points is 10 kHz. Calculate the dynamic impedance of the circuit. (e) Explain how the circuit in (d) is connected to the output of the transistors in (c) to obtain the above conditions. (C&G)

6.7. (*a*) Draw the circuit diagram of a 2-stage, transformer coupled, high-frequency tuned amplifier using transistors in common-emitter configuration. (*b*) Describe the input and output matching and the bias arrangements. (*c*) Explain why the primary winding of each transformer is normally tapped and the secondary circuit is untuned. (*d*) State with reasons a typical value of emitter resistance. (C&G)

6.8. Draw the circuit diagram of a tuned amplifier that employs a transistor connected in the common base configuration. Explain why this type of circuit is sometimes employed and mention the function of each component drawn.

6.9. Draw the circuit diagram of a two-stage r.f. amplifier in which the first stage uses a dual-gate n-channel m.o.s.f.e.t. and the second stage uses an n-p-n bipolar transistor. Outline the operation of your circuit.

6.10. The performance characteristic of an integrated r.f. amplifier is as follows: voltage gain 75 dB, bandwidth 8 MHz–120 MHz, maximum input signal voltage 20 μV, input impedance 12 Ω in parallel with 90 pF. Determine (*a*) the maximum output signal voltage, (*b*) the inductance required to tune the input circuit to 90 MHz if the stray capacitance is 10 pF, (*c*) the inductance required to tune the output circuit to the same frequency assuming the same stray capacitance. (*d*) Draw a possible circuit using this integrated r.f. amplifier and state the purpose of each component shown.

6.11. What is meant by the term *parasitic oscillation* as applied to a tuned amplifier? How may the presence of parasitic oscillations be detected? How may parasitic oscillations be suppressed? Could parasitic oscillations occur in an audio-frequency power amplifier? Give reasons for your answer.

6.12 Draw the circuit diagram of a tuned amplifier that uses the integrated circuit shown in Fig. 6.15. Both the power supply line and the input line connected to pin 3 should be decoupled to earth. Give typical component values and state the purpose of each component.

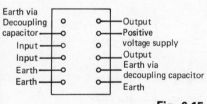

Fig. 6.15

Short Exercises

6.13. List the functions of each of the components shown in Fig. 6.6.

6.14. Draw the circuit of a single-tuned transistor amplifier that uses an r.f. transformer with a tuned primary winding.

6.15. Show, using typical gain/frequency characteristics, how a three-stage amplifier can be stagger tuned.

6.16. Explain how you would measure the output impedance of a tuned amplifier.

6.17. What is meant by the terms sensitivity and selectivity of a tuned amplifier? How is the required selectivity required?

6.18. Draw the circuit of a tuned amplifier that uses a p-channel enhancement-mode m.o.s.f.e.t.

6.19. List the function of each component shown in Fig. 6.16.

6.20. List the function of each component shown in Fig. 6.17.

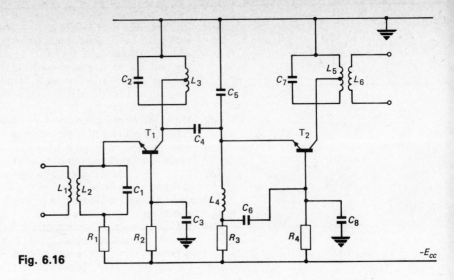

Fig. 6.16

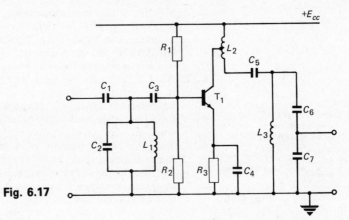

Fig. 6.17

7 Oscillators

Principles of Operation

An oscillator is an electronic circuit designed to produce an alternating e.m.f. of known frequency and waveform. In this chapter only oscillator circuits which produce an output voltage of sinusoidal waveform will be discussed.

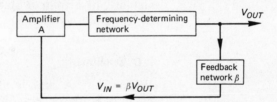

Fig. 7.1 The principle of an oscillator

An oscillator is an amplifier that provides its own input signal, which is derived from the output signal (Fig. 7.1). A fraction β of the output voltage is fed back to the input. If the gain of the amplifier is A_v, the output voltage is $A_v V_{IN}$ and the input voltage is $V_{IN} = \beta A_v V_{IN}$, so that

$$V_{IN}(1 - \beta A_v) = 0$$

The input voltage cannot be zero as an output voltage exists and therefore $(1 - \beta A_v)$ must be zero, or

$$\beta A_v = 1 \tag{7.1a}$$

In general, both the gain A_v and the feedback ratio β are complex and hence

$$|\beta|\,\underline{/\phi} \cdot |A_v|\,\underline{/\theta} = 1\,\underline{/0°} \tag{7.1b}$$

so that

$$|\beta A_v|\,\underline{/\phi + \theta} = 1\,\underline{/0°} \tag{7.1c}$$

127

This equation states the necessary requirements that must be satisfied before a circuit will oscillate: firstly, the loop gain $|\beta A_v|$ must be unity, and secondly, the loop phase shift $(\phi + \theta)$ must be zero.

When an oscillator is first switched on, a current surge in the frequency-determining network produces a voltage, at the required frequency of oscillation, across the network. A fraction of this voltage is fed back to the input terminals of the amplifier and is amplified to reappear across the network. A fraction of this larger voltage is then fed back to the input to be further amplified, and so on. In this way the amplitude of the signal voltage builds up until the gain of the amplifier is reduced in some way to make the loop gain unity. The transistor or f.e.t. may be biased to operate under either Class A or Class C conditions; when the former is used the loop gain is reduced by the transistor driving into saturation. A number of integrated circuit oscillators are also available.

The frequency-determining section of an oscillator may consist of an L-C tuned circuit, a resistance-capacitance network, or a piezo-electric crystal.

The important characteristics of an oscillator are its frequency or frequencies (if variable) of operation, its frequency stability, its amplitude stability, and the percentage distortion of its output waveform.

L-C Oscillators

When Class A bias is used the bias and d.c. stabilization circuitry is the same as that for a transistor amplifier. As the amplitude of the signal increases, the operating point of the transistor is varied over a larger part of its output characteristics until the device is driven into saturation on one half cycle and into cut-off on the other. The gain of the transistor is then reduced to unity as required. The principle is illustrated by the waveforms shown in Fig. 7.2. In Fig. 7.2a the oscillatory voltage applied to the base of the transistor is small and so the peak base current only varies the collector current by a small amount either side of its quiescent value. With increase in the signal amplitude (Fig. 7.2b) the amplitude of the a.c. collector current increases more or less in proportion. With still further increase in the base oscillatory current, the collector current may be varied from approximately zero on one half-cycle to a maximum of E_{cc}/R_L on the other half-cycle. If the input signal amplitude rises still further, one (Fig. 7.2c) or both (Fig. 7.2d) half-cycles of the collector current may be clipped. In Fig. 7.2d the positive half-cycle is clipped because the transistor has driven into saturation and the negative half cycle is clipped since the transistor cuts off. The collector current waveform is

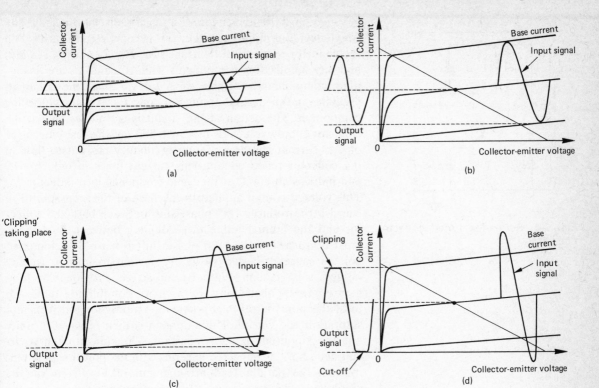

Fig. 7.2 Oscillator transistor driving into saturation and cut-off

non-sinusoidal but this does not lead to a distorted output voltage waveform because the collector load consists of a parallel-tuned circuit of adequate selectivity.

The active device may be a bipolar or a field effect transistor or an operational amplifier. The parallel-tuned circuit may be connected either in series or in parallel with the collector or the drain or, alternatively, may be connected in the base or the gate circuit. The frequency of oscillation is mainly determined by the parallel-tuned circuit and to a good approximation may be taken as equal to the resonant frequency of the circuit.

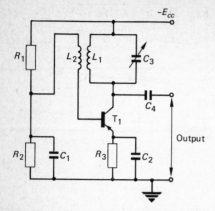

Fig. 7.3 Series-fed tuned collector oscillator

(1) Fig. 7.3 shows the circuit of a *tuned-collector oscillator*; the frequency-determining tuned circuit is connected in series with the collector terminal of the transistor. R_1, R_2 and R_3 are bias and d.c. stabilization components while C_1 and C_2 are decoupling components. Variable capacitor C_3 tunes the circuit to oscillate at the desired frequency and C_4 is a d.c. blocking component. The action of the circuit is as follows. When the collector supply voltage is first switched on, the resulting surge of d.c. current causes a minute oscillatory current to flow in the collector tuned circuit. This current flows in inductor L_1 and induces an e.m.f. at the same frequency into inductor L_2. This voltage is then applied to the base of the transistor. The transistor introduces 180° phase shift between base and collector and the mutual inductance coupling between L_1 and L_2 must be such that the loop phase shift is zero. The amplified voltage causes a larger oscillatory current to flow in L_1 and induces a larger e.m.f. into L_2 and so on. Provided the loop gain is greater than unity the oscillation amplitude builds up until the point is reached where the transistor is driven into saturation and cut-off. The loop gain is then reduced to unity and the amplitude remains constant. The collector tuned circuit need not be connected in series with the power supply but, instead, the parallel-feed arrangement of Fig. 7.4a can be employed. The circuit works in exactly the same way as the series-feed alternative but has the advantage that the one plate of the variable capacitor is at earth, rather than collector supply, potential. The output signal is shown as being taken from a third winding L_3, coupled to L_1 and L_2, but it could be capacitance coupled to the load, as in Fig. 7.3. Other possible circuits are shown in Figs. 7.4b and 7.5. Fig. 7.4b shows a tuned-collector oscillator in which the oscillatory signal is fed into the emitter circuit of the transistor rather than into the base circuit.

Fig. 7.4 Tuned collector oscillators; (a) parallel-fed, (b) using emitter-coupling

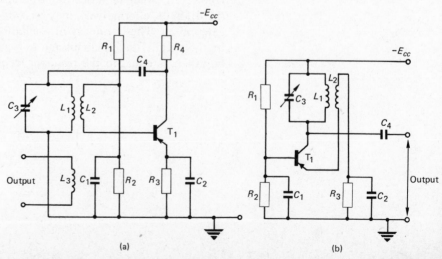

(a)   (b)

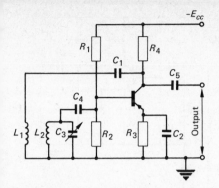

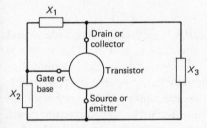

Fig. 7.5 Tuned base oscillator

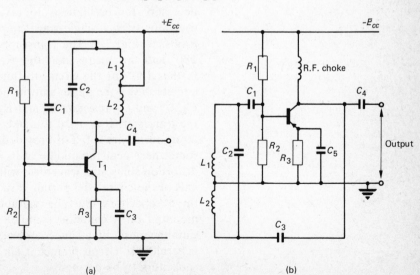

Fig. 7.6 The generalized "three-impedance" oscillator

(2) The circuit of a *tuned-base transistor oscillator* is shown in Fig. 7.5 and can be seen to differ from the parallel-fed tuned collector circuit only in that the inductor L_2 is tuned instead of L_1. A disadvantage of the circuit is that the low input resistance of the transistor considerably reduces the Q-factor of the tuned circuit. This can result in a poor output waveform because insufficient discrimination against harmonics may be provided. The loading effect can, of course, be reduced by connecting the base terminal to a suitable tap on inductor L_2.

(3) Two further types of oscillator which are frequently found in electronic and radio equipment are known as the Hartley and Colpitts oscillators. These oscillators can be represented by the generalized diagram shown in Fig. 7.6, in which X_1 is the reactance connected between base and collector, X_2 is the reactance between base and emitter, and X_3 is the reactance between collector and emitter. For oscillations to occur, reactances X_2 and X_3 must be of the same sign and the reactance X_1 must be of the opposite sign. If X_2 and X_3 are both inductive and X_1 is capacitive, a Hartley oscillator is formed; conversely, for a Colpitts oscillator, X_2 and X_3 are capacitive and X_1 is inductive.

Figs. 7.7a and b respectively show the circuits of the series-fed and parallel-fed HARTLEY OSCILLATORS. In both circuits R_1, R_2 and R_3 provide bias and d.c. stabilization and C_1 is a d.c. block having negligible reactance at the oscillation frequency. Comparing Figs. 7.6 and 7.7 it can be seen that $X_1 = 1/\omega C_2$, $X_2 = \omega L_1$ and $X_3 = \omega L_2$. The frequency of oscillation is given approximately by

$$f_{osc} = \frac{1}{2\pi\sqrt{[C_2(L_1 + L_2)]}} \qquad (7.2)$$

Fig. 7.7 Hartley oscillators

(a)

(b)

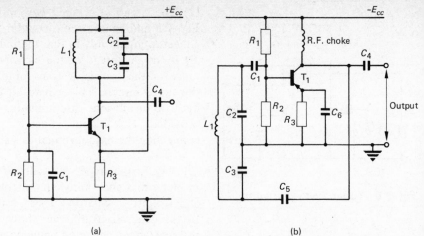

Fig. 7.8 Colpitts oscillators

(a)   (b)

COLPITTS OSCILLATORS employing series-feed and parallel-feed are shown in Fig. 7.8a and b. The required inductive path between the collector and the base terminals of the transistor is collector, inductor L_1, power supply, decoupling capacitor C_1, and base; the reactance of C_1 is negligible at the frequency of oscillation. Similarly, the capacitive path between collector and emitter is via capacitor C_3. The frequency of oscillation is approximately given by

$$f_{osc} = \frac{1}{2\pi\sqrt{L_1\left(\dfrac{C_2 C_3}{C_2 + C_3}\right)}} \qquad (7.3)$$

R-C Oscillators

(1) Audio-frequency oscillators most often use a resistance-capacitance network to obtain the loop phase shift of 360° necessary for oscillations to take place. One type of *R-C* oscillator uses a single-stage amplifier and is shown, in two forms, in Fig. 7.9. The transistor introduces a phase shift of 180° and this means that the *R-C* network must provide a further 180° for the circuit to function as an oscillator. In the circuits shown the phase-shifting network consists of capacitors C_2, C_3 and C_4, resistors R_4 and R_5, and the input resistance of the transistor. It is usual to make $C_2 = C_3 = C_4$ and $R_4 = R_5 =$ input resistance of T_1. Provided the transistor is operated somewhere near the middle of its mutual characteristics, a low distortion sinusoidal waveform will appear at the output terminals of the circuit. To permit variable-frequency operation the three capacitors must be variable and GANGED (that is, mounted on a common spindle) so that they can be varied simultaneously by a single control. Variable-frequency circuits generally use circuit *b* since it allows the moving plates of the capacitors to be earthed.

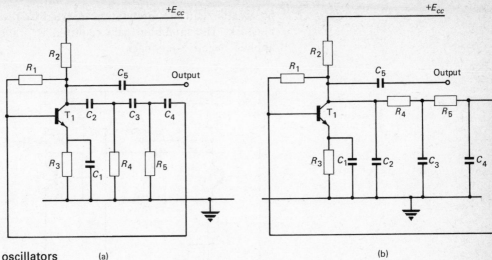

Fig. 7.9 *R-C* oscillators (a) (b)

(2) Another popular version of an *R-C* oscillator is shown in Fig. 7.10 and is known as the WIEN CIRCUIT. A two-stage *R-C* coupled amplifier is used that has an overall phase shift of 360° at the oscillation frequency. A fraction of the output voltage of this amplifier is fed back to its input terminal, the fraction being determined by the *R-C* network R_9-C_3, R_{10}-C_4. The values of R_9, R_{10}, C_3 and C_4 are chosen so that only at the desired frequency of oscillation does the network introduce zero phase shift. This means that at the one particular frequency the phase lead introduced by the series R_9-C_3 circuit is equal to the phase lag given by the parallel R_{10}-C_4 circuit. The loop phase shift of the circuit is then 360° and the circuit will oscillate provided the loop gain is greater than unity. This means that the voltage gain of the amplifier must

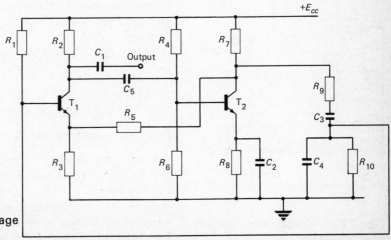

Fig. 7.10 Wien oscillator (voltage feedback)

be greater than the attenuation inserted by the phase-shifting network. The minimum gain required is 3 and is set by the applied negative feedback.

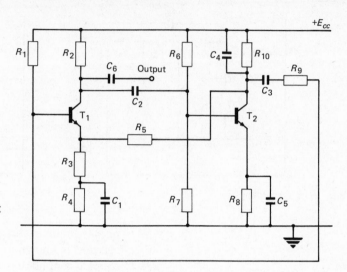

Fig. 7.11 Wien oscillator (current feedback)

The circuit shown in Fig. 7.10 is best suited to use with high-input impedance amplifiers, such as a bipolar transistor stage with voltage-current negative feedback or a f.e.t. stage. When a low-input impedance amplifier is used, the positions of the series and parallel elements of the phase-shifting circuits are best interchanged (see Fig. 7.11). A fraction of the collector current of the second stage is now fed back to the first stage and the current gain of the amplifier must be large enough to maintain oscillations.

The necessary gain can also be provided by an operational amplifier and Fig. 7.12 shows the basic circuit. The operational amplifier is used in the non-inverting mode and for a voltage gain of 3 the resistance of R_2 must be chosen to be twice the resistance of R_1.

For all the Wien circuits the frequency of oscillation is given by

$$f_{osc} = \frac{1}{2\pi\sqrt{(R_9 R_{10} C_3 C_4)}} \tag{7.4}$$

(For the op-amp circuit $R_9 = R_3$, $R_{10} = R_4$, and $C_4 = C_2$.)

Clearly the oscillation frequency is inversely proportional to the values of resistance and capacitance used in the phase-shifting circuit. For this reason R-C oscillators are not used at frequencies much higher than about 50 kHz since stray capacitances would begin to exert a relatively large influence on the frequency of oscillation.

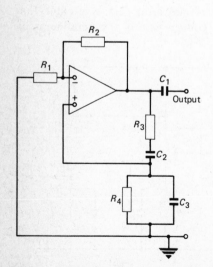

Fig. 7.12 Operational amplifier Wien oscillator

The amplitude of the oscillatory waveform could be limited by allowing the output transistor to saturate, as is often the practice with an L-C oscillator. This would result in a larger percentage harmonic distortion of the output waveform than is normally specified. It is customary, therefore, to employ temperature-dependent negative feedback to provide amplitude stability. This is achieved by replacing the feedback resistor R_5 in Figs. 7.10 and 7.11 and R_2 in Fig. 7.12, by a *negative temperature coefficient* (n.t.c.) resistor, such as a *thermistor*. Overall negative feedback is applied by R_3 and R_5 in the transistor circuits and by R_1 and R_2 in the op-amp circuit. The overall gain of the amplifier is hence very nearly R_5/R_3 or $(R_1 + R_2)/R_1$. As the oscillation amplitude increases, the voltage across R_5 (or R_2) increases also and so does the power dissipated in the component. The temperature of $R_5(R_2)$ increases and its resistance decreases. The decrease in the resistance of $R_5(R_2)$ reduces the gain of the amplifier and returns the output voltage to a stable value.

Frequency Stability

The frequency stability of an oscillator is the amount by which its frequency drifts from the desired value. It is desirable that the shift should be very small and the maximum allowable change in frequency is normally specified as so many parts per million, e.g. ± 1 part in 10^6 would mean a maximum frequency drift of ± 1 Hz if the frequency of oscillation were 1 MHz but ± 100 Hz if the frequency were 100 MHz. The frequency stability of an oscillator may be short-term, minute-by-minute, or long term, hours, days or even longer.

The frequency of an L-C oscillator has hitherto been taken as equal to the resonant frequency of the tuned circuit, and the frequency of an R-C oscillator as a function of the R-C values selected. However, the oscillation frequency also depends upon the load impedance into which the oscillator works and the parameters of the transistor or op-amp.

Load on the Oscillator

The frequency of oscillation is not independent of the load into which the oscillatory power is delivered. If the magnitude of the load should vary, the oscillation frequency will not be stable. Variations in the external load can effectively be removed by inserting a *buffer* amplifier in between the oscillator and the load (see Fig. 7.13). The buffer amplifier serves the dual purpose of isolating the oscillator from any changes in the load and increasing the output voltage level. The amplifier is an ordinary audio- or radio-frequency amplifier, depending on

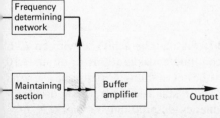

Fig. 7.13 Use of a buffer amplifier

the oscillator frequency. In a radio transmitter the buffer amplifier will be a high-efficiency Class C tuned power amplifier.

Variations in Supply Voltage

The parameters of a transistor, bipolar or f.e.t., such as current gain, mutual conductance, and input and output capacitances, are functions of the quiescent collector or drain current and hence of the supply voltage. Any change in the supply voltage will cause one or more parameters to vary and the oscillation frequency to drift. This cause of frequency instability is fairly small but, if necessary, the stabilization of the power supply voltage can be improved.

Circuit Components

Changes in the temperature of circuit components will produce changes in inductance and capacitance values and thereby change the oscillation frequency. Temperature changes alter the inductance by changing the dimensions of the wire and the former on which it is wound. The capacitance is a function of temperature because the capacitor plates expand or contract slightly, and also because the permittivity of a dielectric is not quite independent of temperature. The frequency stability can be improved by minimizing temperature changes by (a) keeping the power dissipated within the active device small, (b) keeping the frequency-determining components well clear of any heat sources, and (c) (if the expense is justified) mounting the components inside a thermostatically-controlled enclosure or *oven*. Temperature changes can also be minimized by keeping the oscillator permanently switched on. When all temperature changes have been minimized a further improvement in frequency stability can be achieved by using components having small and/or opposite temperature coefficients. The practical difficulty inherent in the use of negative and positive temperature coefficient components is that they must track one another over a range of temperatures and be reproducible in quantity production.

Crystal Oscillators

The best frequency stability that can be achieved with an L-C oscillator is limited by economic considerations to about ± 10 parts in 10^6 per °C, and if better stability is required a *crystal oscillator* must be used. A crystal oscillator is an oscillator circuit in which the frequency-determining network is provided by a *piezoelectric crystal*.

Piezoelectric Crystals

A piezoelectric crystal is a material, such as quartz, having the property that, if subjected to a mechanical stress, a potential difference is developed across it, and if the stress is reversed a p.d. of opposite polarity is developed. Conversely, the application of a potential difference to a piezoelectric crystal causes the crystal to be stressed in a direction depending on the polarity of the applied voltage.

In its natural state quartz crystal is of hexagonal cross-section with pointed ends. If a small, thin plate is cut from a crystal the plate will have a particular natural frequency, and if an alternating voltage at its natural frequency is applied across it, the plate will vibrate vigorously. The natural frequency of a crystal plate depends upon its dimensions, the mode of vibration, and its original position or cut in the crystal. The important characteristics of a particular cut are its natural frequency and its temperature coefficient; one cut, the GT cut, has a negligible temperature coefficient over a temperature range from 0°C to 100°C; another cut, the AT cut, has a temperature coefficient that varies from about +10 p.p.m./°C at 0°C to 0 p.p.m./°C at 40°C and about +20 p.p.m./°C at 90°C. Crystal plates are available with fundamental natural frequencies from about 4 kHz up to about 10 MHz or so. For higher frequencies the required plate thickness is very small and the plate is fragile; however, a crystal can be operated at a harmonic of its fundamental frequency and such *overtone* operation raises the possible upper frequency to about 100 MHz.

The electrical equivalent circuit of a crystal is shown in Fig. 7.14. The inductance L represents the inertia of the mass of the crystal plate when it is vibrating; the capacitance C_1 represents the reciprocal of the stiffness of the crystal plate; and the resistance R represents the frictional losses of the vibrating plate. The capacitance C_2 is the actual capacitance of the crystal (a piezoelectric crystal is an electrical insulator and is mounted between two conducting plates). Since the frictional losses of a crystal plate are small the Q-factor of a crystal is high, and figures of 20 000 or more are easily obtained.

A series-parallel circuit, such as Fig. 7.14, has two resonant frequencies: the resonant frequency of the series arm R-L-C_1, and the parallel resonance produced by C_2 and the effective inductance of the series arm above its (series) resonant frequency. When a crystal is fitted into a circuit, very often a small variable capacitor is connected in series to permit fine tuning of its resonant frequency. The crystal may be operated in its series- or parallel-resonant modes, or may work at a frequency such that it presents an inductive impedance.

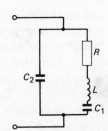

Fig. 7.14 Electrical equivalent circuit of a piezoelectric crystal

Crystal Oscillator Circuits

When the power supply is first connected to a crystal oscil
lator, a voltage pulse is applied to the crystal and causes it t
vibrate at its resonant frequency. An alternating voltage at th
resonant frequency is then developed between the terminals o
the crystal. If this voltage is applied to the base of a transistor
or the gate of a f.e.t., it will be amplified to appear across th
collector or drain load. If some of this amplified voltage i
then, in some way, fed back to the crystal with the correc
phase, it will cause the crystal to vibrate more vigorously. A
larger alternating voltage will appear across the crystal and wil
be amplified and then fed back to the crystal and so on. Th
circuit will therefore oscillate at the frequency at which th
crystal is vibrating.

A number of different crystal oscillator circuits have bee
developed and some examples are given in Fig. 7.15. In th
bipolar transistor circuit of Fig. 7.15a energy is fed from th
collector of T_1 to its base via the crystal X_1. The crystal shoul
have the minimum impedance at the required oscillation fre
quency and so it is operated in its series-resonant mode
Capacitor C_1 provides fine adjustment of the frequency o
oscillation. R_1, R_2, R_3 and C_4 are conventional bias and d.c
stabilization components and the voltage developed across th
tuned circuit L_1-C_2 is coupled to the load (or more likely t
the buffer amplifier) by capacitor C_3.

The circuit shown in Fig. 7.15b is known as the PIERCE
CIRCUIT and is essentially a version of the Colpitts oscil
lator. The required capacitive impedances joining the collecto
to the emitter, and the base to the emitter are provided
respectively, by C_5 and by C_2. The inductive reactance tha
must connect the base to the collector is provided by th
impedance of the crystal in parallel with capacitor C_3. Inducto
L_1 is an r.f. choke.

The f.e.t. oscillator shown in Fig. 7.15c is an example of th
MILLER CIRCUIT. For the circuit to oscillate at a particula
frequency both the crystal and the drain tuned load must have
inductive impedances at that frequency. The necessary feed
back from drain to gate is via the internal electrode capaci
tance of the f.e.t. If the internal capacitance is of insufficien
magnitude it can be augmented by the connection of a
external capacitor of suitable value.

Another version of the Pierce circuit is given in Fig. 7.15d
capacitors C_1 and C_4 provide the capacitive paths betwee
source and gate and source and drain respectively and the
crystal provides the inductive path between drain and gate
Finally, Figs. 7.15e and f show the operational amplifie
versions of a crystal oscillator. Circuit e is a form of Wier

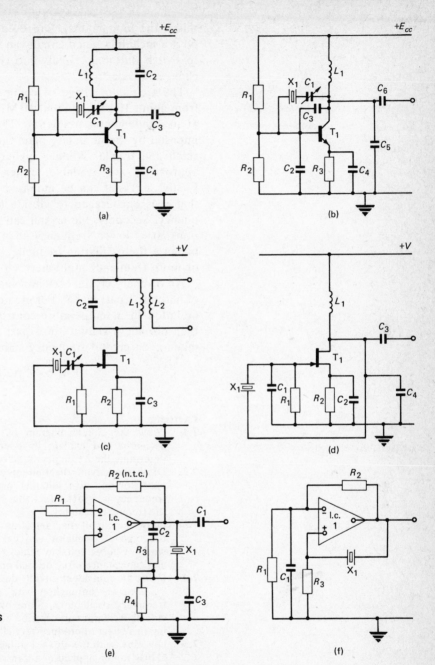

Fig. 7.15 Crystal oscillators

oscillator, while circuit *f* oscillates because positive feedback is applied via the crystal.

When a drain or collector tuned load is employed it is possible to operate a crystal at an *overtone* frequency (i.e. a multiple of the fundamental frequency). Other circuits cannot do this since a crystal will naturally try to work at its most active frequency where its vibrations are most vigorous, and this is the fundamental. On the other hand, oscillators which do not include a tuned circuit can be provided with the facility to switch different crystals into circuit when rapid frequency changes are desired.

The frequency range of commercially available crystals is from about 10 kHz to about 10 MHz with the crystal operating at its fundamental frequency. The higher frequency limit is imposed by the necessary plate thickness becoming extremely small and fragile. When a higher frequency is required two approaches are available, either singly or in combination. Firstly, a crystal can be mounted in its holder in such a way that it is encouraged to vibrate strongly at an *overtone* frequency. Secondly, the crystal can be used to generate oscillations at a lower frequency than that wanted and then to increase the oscillation frequency to the desired value by one or more frequency multipliers.

An ordinary crystal oscillator may have a frequency stability of perhaps 1 part in 10^6. For greater stability the crystal can be mounted in a temperature-controlled environment and may then provide a stability of 1 part in 10^7. If very great care is taken even greater frequency stability can be achieved.

Exercises

7.1. Draw the circuit diagram of a crystal-controlled transistor oscillator and explain its operation. List the factors which determine the frequency stability of such an oscillator. (C&G)

7.2. Draw the circuit diagram of either a Colpitts or a Hartley oscillator and give suitable component values if the required frequency is 1 MHz. List some of the factors which influence the frequency stability. (C&G)

7.3. Draw the circuit diagram of any type of L-C oscillator. Name the type of oscillator you have drawn and describe how it works. Choose suitable values for the tuned circuit inductance and capacitance if the desired operating frequency is 500 kHz.

7.4. Using an equivalent circuit, describe how a quartz crystal can be used in conjunction with a transistor to produce stable frequency oscillations. State one effect of temperature on a crystal oscillator and describe briefly a simple method of reducing this effect to produce very close control. (C&G)

7.5. Describe, with the aid of a suitable waveform sketch, how Class C bias can be applied to a transistor using a leaky base circuit.

7.6. Draw the circuit diagram of an R-C oscillator using an operational amplifier. Describe the operation of the circuit and mention the purpose of each component shown. What are the advantages of using an operational amplifier over the use of a conventional transistor circuit?

7.7. Draw the circuit diagram of (i) a tuned-base oscillator and (ii) a tuned-collector oscillator and explain their operation. State a disadvantage of the tuned-base circuit and suggest how it may be overcome.

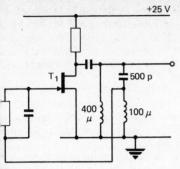

+25 V

T_1

500 p

400 μ

100 μ

Fig. 7.16

7.8. The circuit diagram of a Hartley oscillator is shown in Fig. 7.16. (*a*) Explain its principle of operation. (*b*) Show how the circuit could be modified to form a Colpitts oscillator. (*c*) Upon what factors does the frequency stability of an *L-C* oscillator mainly depend? (*d*) For the circuit given estimate the frequency of oscillation assuming the coils have no mutual inductance and negligible resistance. (C&G)

7.9. Sketch the circuit diagram of a Wien bridge oscillator circuit and describe how oscillations are sustained. State the factors upon which the frequency of oscillation depends. Explain why this type of oscillator must produce harmonic distortion in the output and state how this distortion may be minimized.

(C&G)

7.10. (*a*) Sketch the circuit diagram and describe the operation of one type of tuned oscillator. Identify the section which determines the feedback. (*b*) State the class of bias used and explain why it is not essential to have Class A operation with this type of oscillator. (C&G)

7.11. (*a*) Draw the circuit diagram of a Wien bridge oscillator which uses bipolar transistors. Identify the sections of the circuit which (i) determine the frequency of oscillation (ii) sustain oscillations. (*b*) Briefly explain the operation of the oscillator. (*c*) Give a typical frequency range over which such an oscillator operates. (C&G)

7.12. (*a*) State the conditions necessary to sustain oscillations in a transistor circuit containing a resonant circuit. (*b*) Illustrate your answer by drawing the circuit diagram of a tuned-collector oscillator for use in a superheterodyne receiver. (*c*) State how the output from the oscillator is obtained and the factors that determine (i) the frequency, (ii) the amplitude of the oscillations. (C&G)

7.13. (*a*) State the two main requirements for oscillations. (*b*) With reference to Fig. 7.17, (i) Name the type of circuit, (ii) Describe the operation, (iii) What precautions can be taken to ensure frequency stability? (iv) Why is a high Q desirable in the tuned circuit? (C&G)

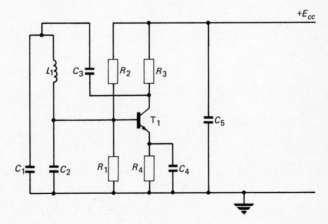

$+E_{cc}$

L_1 C_3 R_2 R_3

T_1

C_5

C_1 C_2 R_1 R_4 C_4

Fig. 7.17

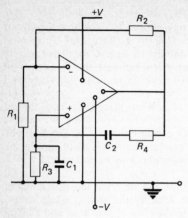

Fig. 7.18

7.14. Refer to Fig. 7.18. (*a*) What is the name and function of th
circuit? (*b*) Suggest a suitable value for resistor R_2. (*c*) Fror
which point in the circuit could an output suitable for operatin
high-impedance headphones be obtained? (*d*) Describe a sim
ple modification to make the output frequency variable. (*e*
What is the type and purpose of C_1?

Short Exercises
7.15. What is meant by Class A and Class C bias of a transistor?
7.16. Describe briefly why an oscillator can be regarded as an amp
lifier with positive feedback applied.
7.17. How is the amplitude of the oscillations limited in (i) an L-(
(ii) an R-C oscillator operating with Class A bias?
7.18. Draw the circuit diagram of an R-C oscillator using an opera
tional amplifier. List the function of each component shown.
7.19. Draw the circuit diagram of a tuned-gate f.e.t. oscillator.
7.20. Draw the circuit diagrams of two different crystal oscillators.
7.21. Draw the circuit diagram of a tuned-drain f.e.t. oscillator. Stat
which type of f.e.t. you have drawn.
7.22. Draw the circuit of a transistor R-C oscillator, assuming th
transistor has a low input resistance.
7.23. Draw the circuit of an R-C oscillator using a field-effec
transistor.
7.24. State the purpose of a buffer amplifier.
7.25. Draw the circuit diagram of a crystal oscillator which uses a\
integrated operational amplifier as its active device. State th
purpose of each component shown.
7.26. Draw the circuit of a tuned-gate oscillator. State which type o
f.e.t. you have drawn.
7.27. Draw the circuit of a Hartley oscillator using a j.f.e.t.
7.28. Draw the circuit of a Colpitts oscillator using an enhancemen
mode m.o.s.f.e.t.

8 Waveform Generators

Introduction

Multivibrators are a class of oscillator that provide an output voltage of rectangular waveform. Positive feedback is applied to a circuit in such a way that the active devices are alternately driven into a saturation and into cut-off. The active devices therefore act as electronic switches and not as amplifiers. Multivibrator circuits can be placed into one of three main categories; namely, the free-running or *astable*, the *monostable*, and the *bistable* or *flip-flop*. An ASTABLE multivibrator operates continuously to provide a rectangular waveform with particular values of pulse repetition frequency and mark/space ratio. A MONOSTABLE circuit has one stable and one unstable state; normally the circuit rests in its stable state but it can be switched over into its alternate state, by the application of an external *trigger* voltage, where it will remain for a certain length of time before returning to its stable condition. Lastly, the BISTABLE circuit has two stable conditions and it will remain in either one until switching is initiated by a trigger signal.

Multivibrator circuits can be designed using bipolar or field-effect transistors, operational amplifiers, logic elements such as NAND and NOR gates [E II], and in some cases are also available as integrated circuits.

Some other kinds of circuit are also commonly used for the generation of non-sinusoidal waveforms. The BLOCKING OSCILLATOR can be used to generate either pulse or ramp waveforms, while the MILLER INTEGRATOR is often used as a ramp or timebase generator. The Schmitt trigger circuit is a circuit that can be used to convert signals of various waveforms, including sinusoidal, into rectangular waveshape.

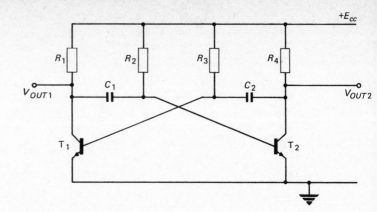

Fig. 8.1 Astable multivibrator

Astable Multivibrators

(1) The circuit diagram of a commonly-employed free-running multivibrator is shown in Fig. 8.1. Resistors R_1 and R_4 are collector load resistors and R_2 and R_3 provide base bias. Capacitors C_1 and C_2 each couple the collector of one transistor to the base of the other transistor. When the collector supply is switched on, both transistors will start to conduct current as base current flows via their respective base resistors. The d.c. current gains and switching times of the two transistors will not be identical and so one of the transistors will conduct a greater current than the other. Suppose the transistor that conducts more readily is T_1. As the collector current of T_1 increases, the voltage developed across its collector load resistor R_1 will rise and cause its collector potential to fall. A negative-going voltage pulse is then passed through capacitor C_1 to reduce the base potential of T_2. This makes T_2 conduct less current. This, in turn, results in the collector potential of T_2 rising towards the collector supply voltage E_{cc} and a positive-going voltage pulse is applied to the base of T_1. T_1 now conducts even harder and its collector voltage falls taking the base potential of T_2 with it. Now T_2 conducts even less current and its collector voltage becomes even more positive causing T_1 to conduct still harder. A cumulative effect takes place in this way which very rapidly leads to T_1 being driven into saturation and T_2 into cut-off. T_2 now has a negative potential approximately equal to E_{cc} volts, relative to earth, at its base terminal, and so capacitor C_1 has its left-hand plate at very nearly zero potential and its right-hand plate at about $-E_{cc}$ volts. C_1 now starts to charge exponentially from $-E_{cc}$ volts to $+E_{cc}$ volts with a time constant of $C_1 R_2$ seconds (see Fig. 8.2).

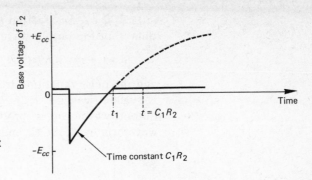

Fig. 8.2 Base voltage of T_2 against time

After a time t_1 the base voltage of T_2 will rise to be slightly more positive than zero volts (actually $+V_{BE2}$ volts) and then T_2 immediately starts to conduct current. The collector potential of T_2 falls and a negative-going voltage pulse is transferred through C_2 to the base of T_1. T_1 now conducts less current and its collector potential rises and the base of T_2 is taken more positive and so on. In this way, T_2 is rapidly turned ON and T_1 is turned OFF. The base of T_1 is now at a potential of approximately $-E_{cc}$ volts but starts rising exponentially, with a time constant C_2R_3, towards $+E_{cc}$ volts. Immediately the base potential of T_1 passes through zero T_1 starts to conduct again and the circuit rapidly reverts to its original state with T_1 ON and T_2 OFF.

This cycle of operation is continuous and results in antiphase rectangular waveforms being available at the two output terminals of the circuit (Fig. 8.3). The time t_1 for which transistor T_1 is non-conducting is given by $t = 0.69C_2R_3$ seconds; simi-

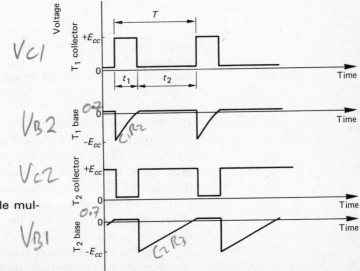

Fig. 8.3 Waveform of astable multivibrator

larly T_1 is OFF for a period $t_2 = 0.69C_1R_2$ seconds. The periodic time T of the output waveform is

$$T = t_1 + t_2 = 0.69(C_1R_2 + C_2R_3)$$

and the pulse repetition frequency is $f = 1/T$ Hz. If the circuit is symmetrical, i.e. $R_2 = R_3$, $R_1 = R_4$, and $C_1 = C_2$, a square output waveform is produced. The periodic time of the waveform is $1.38CR$ and the p.r.f. is $1/1.38CR$ Hz. The

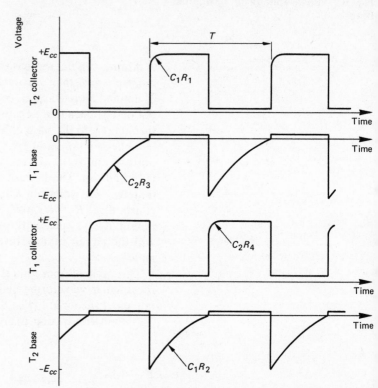

Fig. 8.4 Waveforms of symmetrical astable multivibrator

output waveform of a symmetrical astable multivibrator is shown in Fig. 8.4; it can be see that the leading edges of the output pulses are rounded because of the time constants C_1R_1 and C_2R_4.

(2) The circuit of an astable multivibrator that uses an operational amplifier as the active device is shown in Fig. 8.5. Positive feedback is applied to the op-amp by connecting the junction of the potential divider $(R_2 + R_3)$ across the output terminals of the circuit to the non-inverting$(+)$ terminal. Suppose the output voltage goes positive; a fraction $R_3/(R_2 + R_3)$ of this voltage is fed back to the $+$ terminal and the amplifier is rapidly driven into saturation with positive output voltage $V_{o(SAT)}^+$. The output voltage is also fed back, via resistor R_1, to the inverting $(-)$ terminal and, as this fed back voltage rises,

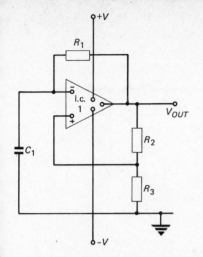

Fig. 8.5 Op-amp astable multivibrator

with a time constant of $C_1 R_1$ seconds capacitor C_1 is charged. After a certain period of time the voltage across C_1 will rise to a value greater than $R_3/(R_2 + R_3)$ times $V_{o(SAT)}^+$. The inverting terminal is now more positive than the non-inverting terminal and the op-amp will immediately switch to provide a negative output voltage. The positive feedback will now ensure that the amplifier very rapidly reaches its negative output voltage $V_{o(SAT)}^-$ condition. Capacitor C_1 now commences to charge up in the opposite direction to before, and immediately the inverting terminal becomes more negative than the non-inverting terminal the circuit will revert to its positive saturated state.

The operational amplifier switches continuously between its positive and negative saturated output voltage conditions with a periodic time T given by

$$T = 2C_1 R_1 \log_e [1 + 2R_3/R_2] \qquad \text{sec} \qquad (8.1)$$

Astable multivibrators are also available in integrated circuit packages from various manufacturers.

Synchronization of an Astable Multivibrator

The frequency stability of an astable multivibrator depends not only on the timing components C_1, C_2, R_2, R_3 but also on the transistor parameters and the stability of the power supplies (as for a sinusoidal oscillator). Very often the frequency stability demanded by a particular application is greater than the inherent stability of the multivibrator. When this is the case an increase in stability to the desired figure can be achieved by inserting a suitable synchronizing signal into the base of one of the transistors.

The synchronizing pulses will be superimposed onto the normal base voltage of that transistor and when the transistor is ON will have no effect. When the transistor is OFF its base voltage is changing exponentially from $-E_{cc}$ volts towards $+E_{cc}$ volts and the transistor switches ON immediately the base voltage reaches zero. The superimposed synchronizing pulses will take the base voltage to zero at a moment in advance of the normal time and will ensure that the transistor always switches ON at the same instant. Fig. 8.6 illustrates the principle; the time taken for the transistor to switch ON is reduced from t_1 to t_2. The multivibrator adjusts its frequency so that the ratio of the frequencies of the synchronizing waveform and the unsynchronized multivibrator is an exact ratio of whole numbers. This ratio can be unity, or greater or less than unity, and is determined by both the amplitude and the frequency of the synchronizing waveform.

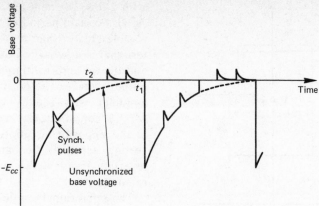

Fig. 8.6 Synchronization of multivibrator

Monostable Multivibrators

(1) The monostable multivibrator is a circuit that has one stable state T_1 ON and T_2 OFF, or vice versa. When the power supply is switched on, the circuit will settle in this stable state and remain there until a *trigger* pulse is applied to initiate switching to the unstable condition. The circuit will then remain in the unstable condition for a time determined by the values of the timing components, before reverting automatically to its stable condition.

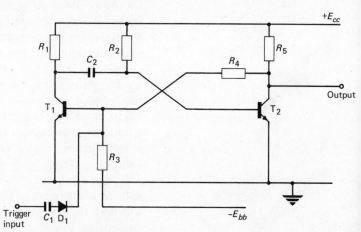

Fig. 8.7 Monostable multivibrator

Fig. 8.7 shows the circuit of a monostable multivibrator. Transistor T_2 is normally conducting because of the base bias current provided by resistor R_2. Transistor T_1 is biased OFF by the negative voltage applied to its base via R_3. The stable condition of the circuit is therefore T_1 OFF and T_2 ON and this means that the output voltage is approximately zero.

If a positive-going voltage pulse is applied to the trigger input terminal, it will be applied via C_1 and D_1 to the base of

T_1 to drive T_1 into its saturated condition. The collector voltage of T_1 falls abruptly from $+E_{cc}$ volts to nearly zero volts and a negative voltage pulse is transferred, via C_2, to the base of T_2. The amplitude of this negative voltage pulse is E_{cc} volts and so T_2 is turned OFF. The collector voltage of T_2 rises suddenly from nearly zero volts to $+E_{cc}$ volts. Capacitor C_2 now has a p.d. of $-E_{cc}$ volts across its plates and starts to charge, with a time constant of C_2R_2, towards $+E_{cc}$ volts (see Fig. 8.2). Immediately the voltage across C_2 and hence the base voltage of T_2, passes through zero volts, T_2 starts to conduct again and its collector potential falls. A negative voltage pulse is applied to the base of T_1, and T_1 conducts less readily and its collector voltage rises positively. A positive voltage pulse is applied to the base of T_2 to make it conduct harder and so on. A regenerative action occurs that rapidly switches the circuit back to its stable state, i.e. T_1 OFF and T_2 ON. Here the circuit remains until another cycle of operation will take place.

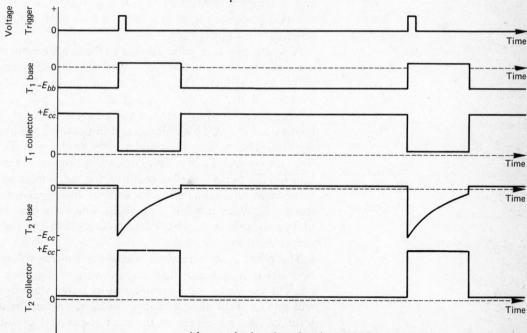

Fig. 8.8 Monostable waveforms

Alternatively, the circuit could be triggered by the application of a negative voltage pulse to the base of T_2 to turn it OFF. Fig. 8.8 shows the waveforms at the various parts of the circuit; rounding off of the collector waveforms (as in Fig. 8.4) has not been shown. The switching speed of the circuit can be increased if the coupling resistor R_4 is shunted by a *speed-up* capacitor of suitable value. The multivibrator will remain in its unstable state for a time t given by

$$t = 0.69C_2R_2 \qquad \text{sec} \tag{8.2}$$

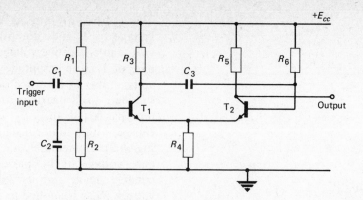

Fig. 8.9 Emitter-coupled mono-stable

The need for a second $(-E_{bb})$ power supply can be avoided if the emitter-coupled circuit of Fig. 8.9 is employed. The base of T_2 is connected to the collector supply line by resistor R_6 whose value is such that T_2 is normally conducting. The emitter current of T_2 flows in the common emitter resistor R_4 and develops a voltage across it that is more positive than the constant positive voltage at which the base of T_1 is held by the potential divider $R_1 + R_2$.

A positive voltage pulse applied to the trigger input terminal increases the base voltage of T_1 to the extent necessary to make T_1 conduct. The collector voltage of T_1 then falls, because of the voltage dropped across R_3, and a negative voltage pulse is transferred via C_3 to the base of T_2. The emitter current of T_2 is reduced and the positive voltage across R_4 falls. This makes T_1 conduct even harder and T_1 rapidly switches ON and T_2 OFF. The circuit will remain in this condition for the period of time required for the voltage across C_3 to become approximately zero volts as the capacitor charges from $-E_{cc}$ volts towards $+E_{cc}$ volts. Once the base potential of T_2 passes through zero, T_2 starts to conduct and the voltage across R_4 rises.

The point is soon reached where the emitter voltage of T_1 becomes more positive than its base voltage and then T_1 cuts off. The circuit has now returned to its stable state in which it will remain until another trigger pulse is applied to its input terminal. An advantage of the emitter-coupled circuit is that the output is taken from a point, the collector of T_2, at which no other connection is made.

(2) The circuit of an operational amplifier monostable multivibrator is given in Fig. 8.10. In the stable condition the potential of the non-inverting terminal is positive with respect to earth by the voltage appearing across resistor R_2. The inverting terminal is at the potential of the negative reference voltage V_{ref}. The output voltage of the circuit is thus at its

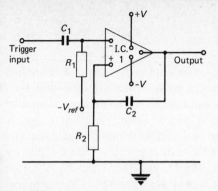

Fig. 8.10 Op-amp monostable multivibrator

positive saturation value. When a positive trigger voltage is applied via C_1 to the inverting terminal, the output of the circuit will rapidly switch over to a negative value as soon as the inverting terminal becomes more positive than the non-inverting terminal. The negative-going change in the output voltage is coupled via C_2 to the non-inverting terminal and drives the operational amplifier into negative saturation. C_2 then discharges towards earth potential at a rate determined by the time constant $C_2 R_2$ seconds. When the voltage at the non-inverting terminal passes the same value as the inverting terminal i.e. $-V_{ref}$, the circuit rapidly reverts to its stable, positive saturated output state. The circuit waveforms are shown in Fig. 8.11. Integrated circuit monostables can be purchased.

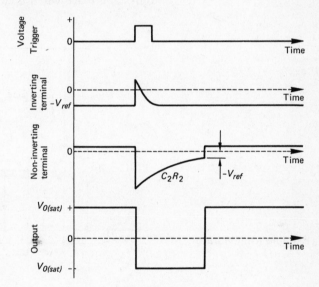

Fig. 8.11 Waveforms in an op-amp monostable

Bistable Multivibrators

(1) The bistable multivibrator, often known as the FLIP-FLOP, is a circuit having two stable states. The circuit will remain in one state or the other until a trigger pulse is applied to cause the circuit to switch to its other state. It will then remain in the second state until another trigger pulse is applied that causes the circuit to revert to its original condition. A number of different types of bistable multivibrator exist and each has its own particular fields of application; these types are known as the SR, the JK, the D, and the T flip-flops. Often flip-flops are *clocked*, that is they are operated in synchronization with a pulse train derived from a crystal oscillator known as the CLOCK. A transition from one state to another can only take place when a clock pulse is obtained.

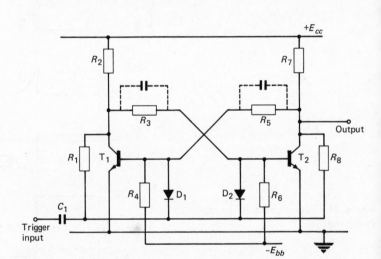

Fig. 8.12 Bistable multivibrator

The circuit diagram of the basic bistable multivibrator is shown in Fig. 8.12. Suppose T_2 is conducting and T_1 is off. The collector voltage of T_2 is then at approximately zero volts, the base-emitter voltage of T_2 is also small, and there is only a small voltage across the diode D_2. Diode D_2 is conducting. The collector voltage of T_1 is at $+E_{cc}$ volts and its base voltage is negative and this means that diode D_1 is reverse biased and non-conducting. If a negative voltage pulse is applied to the trigger input terminal, it will be directed via D_2 to the base of T_2. T_2 will be switched off and its collector potential will rise to $+E_{cc}$ volts and this, in turn, will make the base terminal voltage of T_1 positive. T_1 will now conduct. The circuit has now been switched from one stable condition to another by the application of a trigger pulse.

When T_1 is ON and T_2 is OFF, diode D_1 will be conducting and D_2 will be non-conducting. An input trigger pulse will now

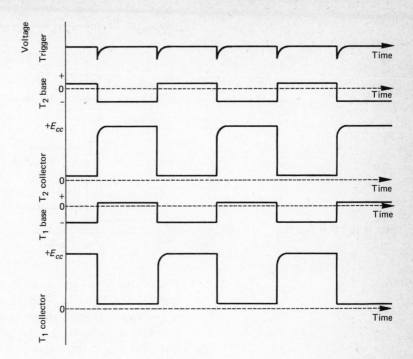

Fig. 8.13 Bistable waveforms

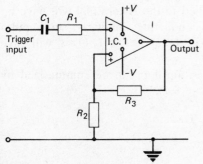

Fig. 8.14 Op-amp bistable multivibrator

be directed to the base of T_1 to switch it OFF and thereby switch T_2 ON. Fig. 8.13 shows the circuit waveforms. Speed-up capacitors may be fitted in parallel with coupling resistors R_3 and R_5 as shown by the dotted connections. The capacitors act by allowing the full instantaneous voltage change to be applied to the base of a transistor.

(2) The bistable multivibrator can also be fabricated using an operational amplifier, the circuit being given in Fig. 8.14. Suppose a positive-going voltage pulse is applied to the trigger input terminal. The pulse is fed, via R_1, to the inverting terminal of the op-amp and a negative voltage appears at the output terminals. A fraction of the output voltage is fed back to the non-inverting terminal as positive feedback and this causes the amplifier to rapidly drive into negative saturation. The circuit will remain in this stable condition until a trigger pulse of opposite polarity is applied and then the circuit switches to its positive saturation condition.

For many applications the need exists for two trigger terminals, known as the set (S) and reset (R) terminals, to allow the circuit to be switched to take up a particular state. The circuit of an S-R flip-flop is shown in Fig. 8.15 Two output terminals are provided; normally one terminal is designated as Q and the other as $\bar{Q}$ (not Q). Suppose the circuit is in the condition where T_1 is cut off and T_2 is saturated, then $Q = 0$ and $\bar{Q} = 1$.

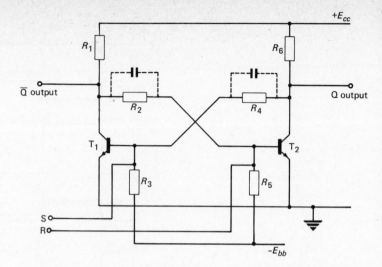

Fig. 8.15 S-R flip-flop

Positive logic is used in which logic 0 corresponds to 0 V and logic 1 to $+E_{cc}$ volts. The circuit will remain in this state until a positive trigger voltage pulse is applied to the S-input terminal of sufficiently large amplitude to drive T_1 into saturation. The fall in the collector voltage of T_1 is transferred to the base of T_2 switching T_2 OFF. The circuit is now in its alternative state with Q = 1 and $\bar{Q}$ = 0. When the trigger pulse ends, T_2 is held OFF by the negative potential derived from the $-E_{bb}$ line and T_1 is held ON by the positive voltage maintained at its base by the high collector voltage of T_2. If, now, a positive-voltage pulse is applied to the R terminal a similar sequence of events will result in T_1 turning OFF and T_2 turning ON; then, once again, Q = 0 and $\bar{Q}$ = 1.

When the circuit is in its first stable state with Q = 0 and $\bar{Q}$ = 1, a positive pulse applied to the R terminal will have no effect since T_2 is already conducting. Similarly, an S pulse applied when the circuit is in the state Q = 1, $\bar{Q}$ = 0 will have zero effect.

The operation of the S-R flip-flop can be summarized by means of a truth table (Table 8.1).

Table 8.1

S	R	Q	Q^+
0	0	0	0
0	0	1	1
1	0	0	1
1	0	1	1
0	1	0	0
0	1	1	0
1	1	0	X
1	1	1	X

Q⁺ represents the state of Q after an S or an R pulse has been applied. If a trigger pulse is simultaneously applied to both the S and the R input terminals, the effect upon the circuit cannot be predicted; the circuit may switch states or it may remain in its existing state.

(3) The S-R flip-flop can also be made by connecting two NOR, or four NAND, gates together in the manner shown in Fig. 8.16. The output of a NOR gate is 1 only when both its inputs are 0. If one or both inputs are 1, the output is 0. Suppose Q = 1 and Q̄ = 0. If S = 1 and R = 0, both inputs to the lower gate will be at 1 and so Q̄ remains at 0. The two inputs to the upper gate are at 0 and Q stays at 1. When S = 0 and R = 1, the upper gate has one input at 0 and one at 1 and hence Q = 0. Both inputs to the lower gate are now 0 and Q̄ becomes 1. Thus the circuit has changed state from Q = 1, Q̄ = 0 to Q = 0, Q̄ = 1. If now S = 1 and R = 0, the lower gate has one input at 1 and the other input at 0. Then Q̄ becomes 0 and so both inputs to the upper gate will be at 0 and Q will switch back to the 1 state. As with the previous circuit the operation of the circuit will be ambiguous if the condition S = 1, R = 1 should arise.

The output of a NAND gate is 0 only when both its inputs are 1, otherwise its output is at logical 1. Suppose that initially Q = 1, Q̄ = S = R = 0; then the inputs to the upper output gate are 1 and 0 and this results in Q = 1. The two inputs to the lower output gate are both 1 and so Q̄ = 0. This is one of the stable states of the circuit. The circuit should change state to Q = 0, Q̄ = 1 if S = 0, R = 1. To confirm this, note that when R becomes 1 the lower output gate will have one input at 1 and the other at 0 and this means that Q̄ will switch to 1. Now the upper output gate has both inputs at 1 and Q becomes 0. This is the other stable state. Again, S = 0, R = 0 will leave the condition of the circuit unchanged while S = 1, R = 1 may result in either output state being obtained.

In some electronic systems it is required to set or reset a flip-flop at particular times determined by *clock* pulses. Synchronized operation of an S-R flip-flop can be achieved by feeding the S and R pulses via AND gates together with the clock pulse train as shown in Fig. 8.17, in which the symbol for the flip-flop is introduced. The output of an AND gate is 1 only when both of its inputs are at 1. This means that when S = 1, R = 0 the flip-flop will not be set until the clock also becomes logic 1, since until this moment the S terminal of the flip-flop itself is not 1. Similarly, when S = 0, R = 1 the circuit will not reset until a clock pulse arrives.

Often the indeterminate S = 1, R = 1 state of an S-R flip-flop cannot be allowed and in such cases an alternative circuit,

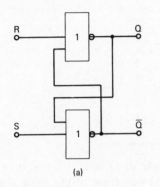

(a)

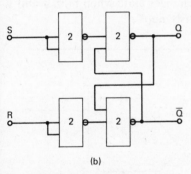

(b)

Fig. 8.16 S-R flip-flop using (a) NOR gates, (b) NAND gates

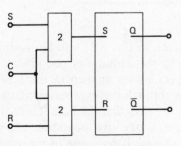

Fig. 8.17 Clocked S-R flip-flop

known as the J-K flip-flop, is employed. The operational difference between the S-R and J-K versions of the flip-flop lies in the final rows of their truth tables. The truth table for a J-K flip-flop is given in Table 8.2.

Table 8.2

J	K	Q	Q^+
0	0	0	0
0	0	1	1
1	0	0	1
1	0	1	1
0	1	0	0
0	1	1	0
1	1	0	1
1	1	1	0

As before, Q^+ represents the state of Q after a J and/or a K pulse has been applied. The final two rows of the table show that the J-K flip-flop always changes state when both J and K trigger pulses are simultaneously applied.

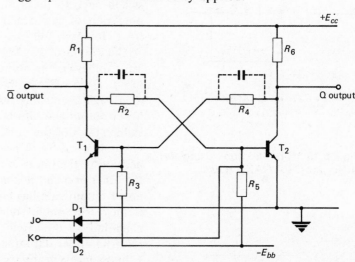

Fig. 8.18 J-K flip-flop

The S-R bistable circuit given in Fig. 8.15 can be modified to perform the J-K function by the addition of two diodes D_1 and D_2 (see Fig. 8.18). Trigger pulses applied to the J or K input terminals will only pass through the appropriate diode if the diode is conducting. This requires that the base terminal of the associated transistor is at a negative potential, i.e. the transistor is non-conducting or OFF. If only one of the trigger terminals has a pulse applied, the circuit will operate in the same way as the S-R flip-flop. If, for instance, T_1 is ON and T_2 is OFF (hence $Q = 1$, $\bar{Q} = 0$) a trigger pulse applied to the J terminal will not pass through diode D_1 and the state of the circuit remains unchanged. On the other hand a pulse applied to the K terminal will pass through diode D_2 and switch transistor T_2 ON, the usual action will then turn T_1 OFF. Now, consider the condition, i.e. $J = K = 1$, that is indeterminate for the S-R flip-flop. When pulses are applied to both inputs simultaneously, only one diode will conduct and allow its pulse to reach the base of the associated transistor. Thus, when simultaneous J and K pulses are applied, only the pulse which will cause the circuit to change state is passed onto the transistors and the operation of the flip-flop is unambiguous. The J-K flip-flop is often clocked.

(4) A third type of bistable is known as the *trigger* or T flip-flop. This is essentially the basic bistable circuit shown in Fig. 8.12 with the addition of a $\bar{Q}$ output terminal. The truth table of a T flip-flop is given by Table 8.3. It is clear that the

Table 8.3

T	Q	Q^+
1	0	1
1	1	0
0	0	0
0	1	1

circuit changes state whenever a pulse is applied to the trigger input terminal. When clocked operation is used, the flip-flop changes state every time a clock pulse is received for as long as the trigger input is 1. The trigger function can be obtained by the use of a J-K flip-flop by merely connecting the J and K terminals together (Fig. 8.19).

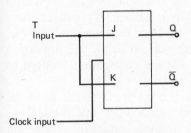

Fig. 8.19 T flip-flop

(5) The fourth type of bistable that is in common use is known as the D flip-flop and its truth table is given in Table 8.4. The D flip-flop has a single trigger input terminal and its Q output terminal always takes up the same logic value as the D input. The circuit is easily derived from an S-R, or a J-K, bistable by merely connecting a NOT stage between the S and R, or J and K, terminals (see Fig. 8.20).

Table 8.4

D	Q	Q^+
1	0	1
1	1	1
0	0	0
0	1	0

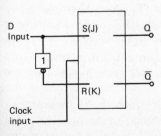

Fig. 8.20 D flip-flop

Integrated circuit versions of the S-R, J-K and D bistable multivibrators are readily available from most, if not all, manufacturers. The T flip-flop is not available in I.C. form since it can easily be obtained from a J-K (or S-R) flip-flop.

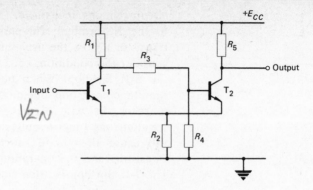

Fig. 8.21 Schmitt trigger

Schmitt Trigger Circuit

The discrete component version of the Schmitt trigger circuit is shown in Fig. 8.21. Two transistors T_1 and T_2 have their emitters connected to a common emitter resistor R_2. A fraction $R_4/(R_3 + R_4)$ of the collector potential of T_1 is applied to the base of T_2. With zero input voltage, T_2 will be fully conducting so that the output voltage of the circuit will be low, approximately 0 V. The emitter current of T_2 will develop a positive voltage across R_2 so that the base/emitter voltage of T_1 will be negative. Hence T_1 is OFF.

If a positive voltage is applied to the input terminals of the circuit, T_1 will start to conduct immediately this voltage becomes more positive than the voltage across R_2 plus about 0.5 V. When T_1 conducts, its collector voltage falls and a fraction of this fall is transfered to the base of T_2. T_2 will therefore conduct less current and the voltage across R_2 will become less positive. This, in turn, increases the forward bias applied to T_1 and so this transistor conducts harder. This is a form of *positive feedback* and, provided the loop gain is greater than unity, will result in T_1 turning ON and T_2 turning OFF. The output voltage of the circuit is then high at E_{cc} volts.

Once the circuit is in the state T_1 ON and T_2 OFF, it will remain in it until the input voltage is reduced to the value at which T_1 comes out of saturation. Then the collector current of T_1 falls and the collector voltage of T_1 increases. The base voltage of T_2 becomes more positive and at some point T_2 starts to conduct. The emitter of T_1 then becomes more positive than its base and T_1 turns OFF. The collector voltage of T_1 rises to $+E_{cc}$ volts and the base voltage of T_2 to $E_{cc}R_4/(R_3 + R_4)$ and this is more than enough to turn T_2 ON. The output voltage of the circuit is then once again at its low value.

The operation of the circuit is illustrated by the waveforms given in Fig. 8.22a. The difference between the positive-going

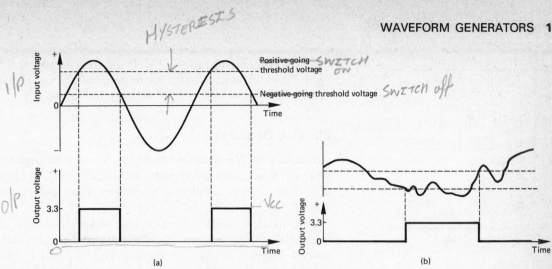

Fig. **8.22** Waveforms in (a) a Schmitt trigger circuit and (b) a NAND Schmitt trigger circuit

and the negative-going *threshold voltages* is known as the *hysteresis* of the circuit.

The Schmitt trigger circuit is used to convert signals of varying waveshape into rectangular pulses and has many applications. Two of these are the marking of the instants in time when a voltage reaches some particular level, and the improvement in the waveform of a rectangular signal that has suffered some attenuation and/or distortion.

The Schmitt trigger circuit can be made using an op-amp (see Fig. 8.23). The reference voltage is chosen to give the required threshold voltages and it can be zero. The operation of the circuit is left as an exercise (8.17). An inverting version of the circuit is obtained by merely interchanging the terminals to which the input and reference voltages are applied.

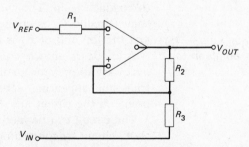

Fig. **8.23** Op-amp Schmitt trigger

Some integrated circuit Schmitt triggers, e.g. the cmos 4093 or the ttl 7413, are associated with an input NAND gate. The input to the trigger part of the device will be high only when all the inputs to the NAND gate are low. Suppose, for example, that the waveform shown in Fig. 8.22*b* is applied to the commoned inputs of the NAND gate. The output of the trigger will then remain at 0 V until the input voltage reaches the lower threshold voltage. At this moment the output vol-

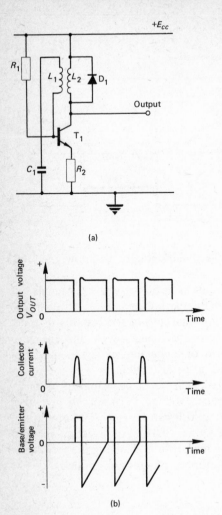

(a)

(b)

Fig. 8.24 (a) Blocking oscillator; (b) waveforms at various points in (a)

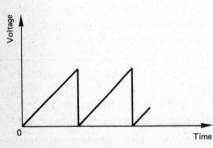

Fig. 8.25 A sawtooth waveform

tage will suddenly rise to +3.3 V. It will then remain at +3.3 V until the input voltage becomes more positive than the upper threshold voltage and then it will abruptly fall to 0 V.

Blocking Oscillator

A blocking oscillator, shown in Fig. 8.24a, can be used for the same purposes as an astable or a monostable multivibrator but it is more usually employed when very short pulses are required.

The circuit is similar in appearance to some versions of *LC* sinusoidal oscillators but it differs from them in that the mutual inductance between the two coils L_1 and L_2 is very much higher. The tight coupling between the coils is obtained by the use of an iron-cored or a ferrite-cored transformer.

When the power supply is first switched on, a current flows into the base of T_1 and a collector current starts to flow. An e.m.f. is induced into L_1 that drives an increased base current into T_1 and so its collector current also increases. This, of course, further increases the e.m.f. induced into L_1 and so on. The transistor *very* rapidly drives into its saturated condition. The collector current is now more or less constant and so the e.m.f. induced into L_1 falls to zero. The base current, and hence the collector current, of T_1 falls and an e.m.f. of the *opposite polarity* is induced into L_1. This makes both the base and the collector currents rapidly fall to zero and charges capacitor C_1 to have a negative voltage with respect to earth.

C_1 now commences to charge towards +E_{cc} volts and, when its voltage becomes more positive than about +0.5 V, T_1 starts to conduct again. The cycle of operation is then repeated. In some circuits C_1 is connected across the emitter resistor R_2 and the top end of L_1 is taken directly to earth.

A diode is shown connected in parallel with L_2; this is necessary to prevent a voltage sufficiently large to damage the transistor appearing across L_2. The waveforms at various points in the circuit are shown in Fig. 8.24b.

The circuit can be modified to operate as a triggered "one-shot" circuit by connecting R_1 to earth instead of to +E_{cc}. The trigger pulses are generally applied to the base of T_1 to turn the transistor ON.

Sawtooth Generators

The sawtooth waveform (Fig. 8.25) is a voltage that rises linearly with time, known as a *ramp*, which when it reaches its maximum value falls rapidly to zero. Immediately the voltage has fallen to zero, it starts another ramp and so on.

A large number of circuits have been devised to act as

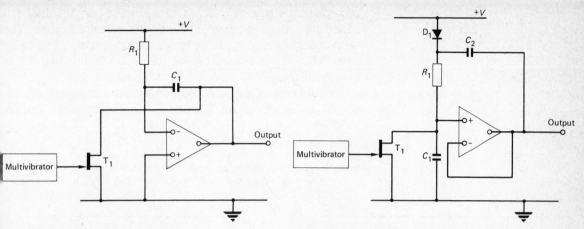

Fig. 8.26 Op-amp Miller integrator sawtooth generator

Fig. 8.27 Op-amp bootstrap sawtooth generator

sawtooth generators but the two most popular are known as the *Miller Integrator* and the *Bootstrap* generator.

The op-amp version of the MILLER INTEGRATOR is shown by Fig. 8.26. When a free-running generator is required, the multivibrator is of the astable type; for a triggered sawtooth generator a monostable multivibrator would be used. The purpose of the multivibrator is to turn the j.f.e.t. ON and OFF.

When T_1 is ON, the voltage across C_1 is very nearly $0\,V$ because the ON resistance of T_1 is very much smaller than R_1. When T_1 is OFF, capacitor C_1 charges up and the voltage across it appears at the inverting terminal of the op-amp and is amplified to provide an output ramp waveform. The ramp waveform produced is very linear because C_1 only charges up a small amount before T_1 turns ON to end the ramp. This means that only the initial linear part of the voltage/time characteristic of a charging capacitor is used.

The circuit of an op-amp BOOTSTRAP SAWTOOTH GENERATOR is shown in Fig. 8.27. C_2 is the *bootstrap capacitor* and it is usually several μF in value. The op-amp is connected as a voltage follower and so it has a voltage gain of unity. When T_1 is ON, the voltage at the non-inverting terminal is $0\,V$. D_1 is ON and, assuming its ON resistance to be very small, there is approximately V volts across both R_1 and C_2.

When T_1 is OFF, C_1 charges via D_1 and R_1. The voltage at the output terminals follows the capacitor voltage and the *change* in voltage is passed, via C_2, to the top of R_1. The voltage at the top of R_1 is then equal to $+(V+v_c)$ and this turns diode D_1 OFF. Now R_1 has a voltage of $(V+v_c)-v_c$ or V across it and *this voltage remains constant* as C_1 charges up. This means that a constant charging current of V/R_1 is provided to charge C_1. Since $Q = It = V_cC$ or $V_c = It/C$ the capacitor voltage increases *linearly* with increase in time. The output voltage waveform is hence a linear ramp. The ramp ends when T_1 is turned ON by the multivibrator and C_1 is rapidly discharged.

The discrete component version of the circuit merely replaces the op-amp with a transistor connected as an emitter follower.

Exercises

8.1. Explain the operation of the circuit shown in Fig. 8.28. Determine its frequency of operation and sketch the output

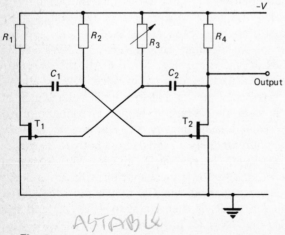

Fig. 8.28

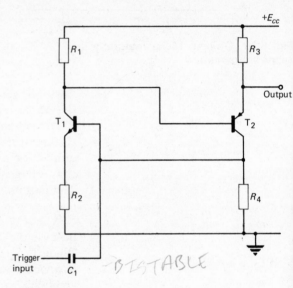

Fig. 8.29

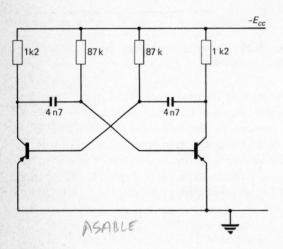

Fig. 8.30

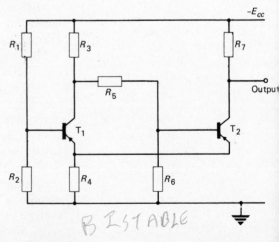

Fig. 8.31

waveform.

8.2. Fig. 8.29 shows the circuit of a bistable multivibrator that uses a complementary pair of transistors. Explain the operation of the circuit.

8.3. Explain the operation of the astable multivibrator shown in Fig. 8.30. Explain why the use of j.f.e.t.'s instead of bipolar transistors allows a much wider range of frequencies to be covered.

8.4. Fig. 8.31 shows the circuit of an emitter-coupled bistable multivibrator. Explain the operation of the circuit.

8.5. With the aid of truth tables explain the differences between the following types of bistable multivibrator: (i) S-R, (ii) J-K, (iii) T, and (iv) D.

8.6. List the various ways in which a monostable multivibrator can be fabricated. Discuss the relative merits of the methods listed. Why is it necessary to trigger a monostable circuit?

8.7. With the aid of a truth table and a circuit diagram explain the operation of a bistable multivibrator. Explain how the circuit is triggered and give waveforms at significant points in the circuit.

Short Exercises

8.8. State the requirements of a transistor multivibrator.

8.9. State the essential differences between the three types of multivibrator.

8.10. What is meant by the term *synchronization* when applied to an astable multivibrator?

8.11. Why are (i) astable multivibrators not triggered, and (ii) bistable multivibrators not synchronized?

8.12. With the aid of a diagram explain how an operational amplifier may be connected to form a multivibrator. State which type you have drawn.

8.13. State the factors which influence the switching speed of a multivibrator. State methods of increasing the speed.

8.14. Explain how two NAND gates can be connected to form an S-R flip-flop. How can the circuit be clocked?

8.15. Use a truth table to explain the difference between the S-R and J-K flip-flops.

8.16. Use a truth table to explain the difference between a D and a T flip-flop.

8.17. Explain the operation of the op-amp Schmitt trigger circuit shown in Fig. 8.23.

8.18. Draw clearly the circuit of a bipolar transistor Miller Integrator sawtooth generator. Explain its operation.

8.19. Draw and explain the operation of a bipolar transistor bootstrap sawtooth generator.

9 Noise

Introduction

The output of a communication system, be it line or radio, will always contain some unwanted voltages or currents in addition to the desired signal. The unwanted output signal is known as *noise* and may have one or more of a number of different causes, each of which will be discussed in this chapter. For the signal received at the end of a system to be of use, the signal power must be greater than the noise power by an amount depending upon the nature of the signal. The ratio of the wanted signal power to the unwanted noise power is known as the SIGNAL-TO-NOISE RATIO. The signal level must never be allowed to fall below the value that gives the required minimum signal-to-noise ratio, because any gain introduced thereafter will increase the level of both noise and signal by the same amount and will not improve the signal-to-noise ratio. Noise having a constant energy per unit bandwidth over a particular frequency band is said to be WHITE NOISE.

Thermal Agitation Noise

If the temperature of a conductor is increased from absolute zero (−273°C), the atoms of the conductor begin to vibrate and some electrons are able to break away from their parent atoms and wander freely within the conductor. The amplitude of the atomic vibrations increases with increase in the temperature of the conductor and so the number of free electrons also increases with temperature. The free electrons wander in a random manner within the conductor, but at any particular instant more electrons are travelling in some directions than in others. The movement of an electron constitutes the flow of a minute current, and therefore the net current represented by the moving electrons fluctuates continuously in both mag nitude and direction. Over a period that is long compared with the average time an electron travels in a particular direction, the total current is zero. The continuous flow of minute currents develops a random voltage across the conductor, and this unwanted voltage is known as *thermal agitation* noise or *resistance* noise.

It can be shown that the r.m.s. noise voltage produced by thermal agitation in a conductor is given by

$$V_n = \sqrt{(4kTBR)} \tag{9.1}$$

where k = Boltzmann's constant = 1.38×10^{-23} J/°
$\quad T$ = temperature of conductor
$\quad B$ = bandwidth (Hz) over which noise is measured, or of circuit at whose output the noise appears, whichever is the smaller. For most practical purposes, B can be taken as the 3 dB bandwidth of the circuit
$\quad R$ = resistance of circuit, ohms

Equation (9.1) may also be extended to find the noise voltage produced in an impedance; R is then the resistive component.

It is the bandwidth and not the frequency of operation that is important with regard to thermal agitation noise. Thus a wideband amplifier is noisier than a narrow-band amplifier whatever their operating frequencies may be.

EXAMPLE 9.1

Calculate the noise voltage produced in a 10 kΩ resistance in a 1 MHz bandwidth if the temperature is 20°C.

Solution
From equation (9.1),

$$V_n = \sqrt{(4 \times 1.38 \times 10^{-23} \times 293 \times 1 \times 10^6 \times 1 \times 10^4)} = 12.7 \ \mu\text{V} \qquad (Ans.)$$

EXAMPLE 9.2

A parallel resonant circuit has an inductance of 5 μH and Q-factor of 50. If the circuit is resonant at a frequency of 10 MHz, calculate the thermal noise generated. Assume a temperature of 300°.

Solution
The thermal agitation noise is generated in the resistive component of the impedance of the tuned circuit. At the resonant frequency of 10 MHz this is the dynamic resistance $R_d = Q\omega_0 L$, i.e.

$$R_d = 50 \times 2\pi \times 10^7 \times 5 \times 10^{-6} = 5\pi \ \text{k}\Omega$$

The bandwidth is the 3 dB bandwidth of the circuit:

$$B_{3\text{dB}} = \frac{f_0}{Q} = \frac{10^7}{50} = 200 \ \text{kHz}$$

Thus, from equation (9.1),

$$V_n = \sqrt{(4 \times 1.38 \times 10^{-23} \times 300 \times 200 \times 10^3 \times 5\pi \times 10^3)}$$
$$= 7.22 \ \mu\text{V} \qquad (Ans.)$$

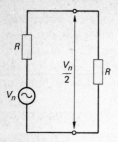

Fig. 9.1   Available noise power

The thermal noise e.m.f. may be regarded as acting in series with the resistance R producing it. Maximum power transfer from a resistive source to a load occurs when the load resistance is equal to the resistance of the source. Consider a resistance R connected across another resistance of the same value that may be considered to be noiseless (Fig. 9.1). The noise power delivered to the load resistance is

$$P_a = \frac{(V_n/2)^2}{R} = \frac{4kTBR}{4R} = kTB \text{ watts} \qquad (9.2)$$

Thus the maximum or AVAILABLE NOISE POWER that can be delivered by a resistance is independent of the value of that resistance but is proportional to *both* temperature *and* bandwidth. It is often convenient to note that, if the temperature is 290 K (17°C), the available noise power is 4×10^{-15} W/MHz. Thermal agitation noise is *white*.

EXAMPLE 9.3

Calculate the available noise power from a resistance at a temperature of 17°C over a 2 MHz bandwidth.

$$P_a = 8 \times 10^{-15} \text{ W} \qquad (Ans.)$$

The noise produced in most metallic resistors is purely the thermal agitation noise given by equation (9.1). Most carbon resistors produce CURRENT NOISE in addition to this; current noise is caused by random variations in the contact resistance between carbon particles. The current noise e.m.f. increases with increase in both the current flowing in the resistor and in the resistance value, and is inversely proportional to frequency. At audio frequencies current noise may be larger than thermal agitation noise, but at radio frequencies thermal agitation noise predominates. The total noise voltage V_{tn} produced in a carbon resistance is given by

$$V_{tn} = \sqrt{[(\text{thermal noise})^2 + (\text{current noise})^2]} \qquad (9.3)$$

Noise in Semiconductors

So far in this book the output current of a transistor, f.e.t. or integrated circuit has been taken as comprising the direct current determined by the d.c. operating conditions, plus a superimposed alternating current that is proportional to the input signal current or voltage. Random fluctuations in these currents always exist, however, and may be considered to be the result of the superimposition of a noise current on the direct and signal currents.

Noise in Transistors

THERMAL AGITATION NOISE. With zero voltages applied to the terminals of a transistor, charge carriers, both electrons and holes, move randomly within the semiconductor material. When direct voltages are applied the random motion is superimposed on a general current flow from positive to negative. A thermal noise voltage is developed in the base spreading resistance r_b of the transistor, given by $\sqrt{(4kTBr_b)}$. The same effect occurs in the emitter and collector regions also, but is negligible in comparison with the noise in the base region.

SHOT NOISE. Shot noise in a transistor is caused by the random arrival and departure of charge carriers by diffusion across a p-n junction. Since there are two p-n junctions in a transistor there are two sources of shot noise.

PARTITION NOISE. The input current to a transistor flows through the emitter to the base–emitter junction. After crossing the junction it divides between the collector and base terminals ($I_E = I_B + I_C$). This current division is also subject to random fluctuations and is thus another source of noise.

$1/f$ NOISE. Fluctuations in the conductivity of the semiconductor material produce a noise source which is inversely proportional to frequency. This noise, also known as current or excess noise, is usually negligible above about 10 kHz, and for some transistors above about 1 kHz.

Noise in F.E.T.s

Noise in an f.e.t. originates from three sources: shot noise generated by leakage currents in the gate-source p-n junction, thermal noise generated in the channel resistance, and $1/f$ noise caused by the random generation and recombination of charge carriers. The f.e.t. is inherently a lower noise device than is a transistor, although if a transistor is operated with a collector current of only a few microamperes its noise performance may be superior.

The reasons for the generally superior noise performance of a field-effect transistor are: (i) that its structure contains only one p-n junction as opposed to two in a bipolar transistor; this means that shot noise is less; and (ii) that the current flowing into the source can only flow out of the drain and this means that the f.e.t. is not subject to partition noise.

Noise in Line Systems

The noise output of a line communication system consists of noise generated in the transmission line itself and noise produced within the repeater stations along the route.

The noise arising in the transmission line is the sum of thermal agitation noise in the line resistance, noise due to faulty joints, interference picked up from nearby power lines or electric railways, and crosstalk from other pairs in the same cable. Thermal agitation noise has already been considered and noise caused by a faulty joint needs no discussion.

If a transmission line runs more or less parallel to a power line or an electric railway, it may have unwanted power-frequency voltages induced in it via inductive and/or capacitive couplings between the lines. Underground cables often have a metallic sheath, and this acts as a screen to reduce the magnitude of the unwanted voltages. Coaxial pairs are generally operated with their outer conductor earthed and are quite efficiently self-screened. This type of interference is minimized by keeping telecommunication cables spaced as far away from power lines as possible.

CROSSTALK is a voltage appearing in one pair in a cable when a signal is applied to another pair. Any multi-pair cable will experience crosstalk between all its pairs to a greater or less extent. Crosstalk in a cable is caused by electrical couplings between the conductors; these couplings may be capacitive, magnetic or via insulation resistances. The construction of a cable is designed to minimize crosstalk, and, when necessary, balancing the couplings between pairs at the end of each section of line can give a further reduction.

In a repeater station, noise voltages are introduced by thermal agitation in the equipment, faulty connections, inadequate smoothing of the power supplies, i.c., transistor and f.e.t. noise, and crosstalk. Crosstalk occurs in the internal wiring because of capacitive and inductive couplings between wires. In the electronic equipment crosstalk is the result of electric couplings between inadequately screened components and couplings via the common power supplies. Crosstalk via the power supplies can occur because the current taken by each circuit flows in the common internal impedance of the power supply and develops a voltage across it. To minimize crosstalk in power supplies they are designed to have very low impedance.

A major cause of noise in multi-channel systems is known as INTERMODULATION NOISE. If a complex wave is applied to a device having a non-linear input/output characteristic, a number of new frequencies are produced and are present at the output. These new frequencies are equal to the sums and/or differences of the frequencies contained in the input signal. For example, if the input signal has components at frequencies f_1 and f_2 the output signal will contain components at

$$f_1 \pm f_2, \; 2f_1 \pm f_2, \; 2f_2 \pm f_1, \; 3f_1 \pm 2f_2, \text{ etc.}$$

in addition to the original frequencies f_1 and f_2. The number of extra frequencies thus produced can be very large; for example, if the input signal contains 100 different frequencies the number of *intermodulation products* runs into millions. In a 12-channel telephony system [TS II] with all channels transmitting speech signals, the intermodulation products are beyond count and produce intermodulation noise at the output of the system. Intermodulation noise has components at all frequencies within the bandwidth of a channel and sounds very similar to thermal agitation noise. However, whereas thermal agitation noise is continuously produced, intermodulation noise depends upon the amplitude of the signals applied to the channels of the system.

Noise in Radio Systems

The output signal of a radio system is always accompanied by noise generated in the radio equipment and picked up by the receiving aerial. The sources of noise generated in the radio equipment itself have been previously discussed: thermal agitation noise, i.c., transistor and f.e.t. noise, and in some cases intermodulation noise. Noise picked up by an aerial may be either man-made or natural.

Man-made Noise

(1) The available frequency spectrum for the various radio services is limited, and the bandwidth allocated to each channel, or station, is the minimum practicable. For example, medium-wave broadcast stations have a bandwidth of about 9 kHz. One consequence of this is that signals proper to transmissions at frequencies close to that of the wanted signal may also be received. Such *adjacent-channel* signals cause interference and/or whistles at the output of a receiver, and this interference may be considered to be a form of noise. ADJACENT-CHANNEL INTERFERENCE is reduced by arranging that transmitters operating at the same or adjacent frequencies are as widely spaced, geographically, as possible, and by using receivers of adequate selectivity.

(2) When an electric current is switched on and/or off its waveform is abruptly changed and a number of components at radio frequencies are produced. These radio-frequency components may be radiated directly from the point where the current is interrupted, or they may be propagated along the mains wiring and then radiated from a distant point. Mains radiation accounts for most cases of interference since the interference can be radiated over a fairly wide area. Typical sources of such interference are electric motors in domestic apparatus, neon signs and car ignition systems. This form of interference may be significant up to 100 MHz or so.

The obvious way of overcoming INTERFERENCE FROM ELECTRICAL APPARATUS is to position the receiving aerial well distant from all possible sources of interference and to use a screened lead from the aerial to the receiver. This is generally possible at the receiving ends of point-to-point links but quite impracticable for domestic broadcasting. To reduce the interference with broadcasting to tolerable proportions it is therefore necessary to suppress the interference at the source. An electric motor, for example, will produce interference if there is sparking at its brushes. The interference may be propagated into the mains wiring by one of two modes, symmetrical or asymmetrical, indicated by arrows in Fig. 9.2. Symmetrical interference can be reduced either by increasing the r.f. impedance of the mains by inserting a series inductor in each line (Fig. 9.3*a*), by reducing the impedance of the mains with shunted capacitors, as at Fig. 9.3*b*, or by a combination of both methods. Suppression of asymmetrical interference is more complex, and one possible arrangement is shown in Fig. 9.4. The r.f. interference currents have two alternative paths: either a high-impedance inductive path to the mains or a low-impedance capacitive path. The r.f. currents flowing in the high-impedance path pass through one winding of the transformer and induce an e.m.f. in the other winding. It is easy to deduce that the directions of the induced e.m.f.s are such as to still further reduce the interference currents that reach the mains.

Car ignition systems rely on the constant interruption of a low-voltage circuit to obtain the high voltage necessary to produce sparks in the engine (Fig. 9.5). Current flowing in the low-voltage circuit is periodically interrupted by a cam-operated contact breaker, driven from the engine. A high voltage is thus induced into the other winding of the coil. At the same time a rotor arm in the distributor connects the secondary winding of the coil to one of four electrodes, each of which is connected to a sparking plug by a lead. The arrangement is carefully timed so that a spark is produced at each cylinder in the engine at the optimum time for igniting the petrol-air mixture.

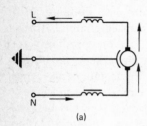

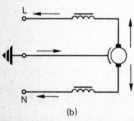

Fig. 9.2 Symmetrical and non-symmetrical modes of propagation of r.f. interference into a mains supply

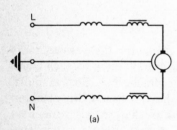

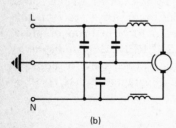

Fig. 9.3 Suppression of symmetrical interference

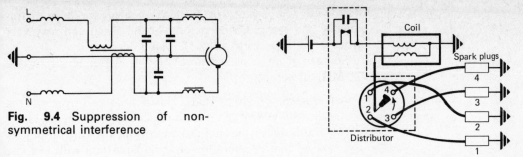

Fig. 9.4 Suppression of non-symmetrical interference

Fig. 9.5 Car ignition system

Adequate suppression of r.f. currents is obtained by the use of high-resistance, screened leads from coil to distributor and from distributor to sparking plugs.

Natural Noise

At frequencies up to about 20 MHz the predominant naturally produced noise is ATMOSPHERIC or STATIC NOISE. Flashes of lightning or other electrical discharges in the atmosphere produce signals that are picked up by an aerial and give rise to noise at the receiver output. When the discharges are intermittent, noise impulses are heard, but when a large number of discharges occur in a short period the result is very similar to thermal-agitation noise. Atmospheric noise can be propagated through the ionosphere over long distances, and since electrical discharges are normally occurring somewhere in the world, continual static is present.

Above 20 MHz or so the total noise picked up by an aerial becomes increasingly thermal in nature and may be given by

$$V_a = \sqrt{(4ktT_0BR_r)} \tag{9.4}$$

where T_0 is the ambient temperature (normally taken as 290°), t is the NOISE TEMPERATURE of the aerial, and R_r is the radiation resistance of the aerial. B and k have the same meaning as before.

An aerial must be considered to be at a temperature other than its actual temperature if equation (9.4) is to give the correct noise voltage. This is because the aerial picks up noise from a number of different sources, each of which will be mentioned shortly. The *noise temperature* of an aerial is the temperature at which its radiation resistance must be considered to be in order to produce the same noise as is actually picked up, i.e.

$$t = \frac{\text{Total aerial noise power}}{kT_0B} \tag{9.5}$$

The noise temperature of an aerial depends upon both frequency and the radiation pattern of the aerial and may vary from a few degrees to a few thousand degrees.

Sources of noise picked up by aerials at frequencies above 20 MHz or so are:

(a) *Galactic or cosmic noise*: this is produced from stars in the Milky Way and it is prominent up to about 1 GHz.

(b) *Earth noise*: thermal radiation from the earth produces noise that becomes increasingly important with increase in frequency above approximately 200 MHz.

(c) Above 10 GHz *quantum noise* becomes increasingly significant; this is noise caused by changes in the atomic and molecular states of particles in the atmosphere.

(d) Noise produced by radiation from the sun may be important between about 100 MHz to 10 GHz or so.

Signal-to-Noise Ratio

The lower limit to the signal level that can be usefully amplified is set by the noise which is unavoidably present with the signal. The ratio of the wanted signal *power* to the unwanted noise *power* is known as the signal-to-noise ratio, i.e.

$$\text{Signal-to-noise ratio} = \frac{\text{Wanted signal power}}{\text{Unwanted noise power}} \quad (9.6a)$$

or

$$\text{Signal-to-noise ratio}$$
$$= 10 \log_{10}\left(\frac{\text{Wanted signal power}}{\text{Unwanted noise power}}\right) \text{decibels} \quad (9.6b)$$

The signal-to-noise ratio required of a particular system depends upon the potential use of the signal and is generally determined by means of subjective tests. For example, a line music circuit may require a signal-to-noise ratio of 60 dB in order that the transmitted music may not be noticeably degraded, but a low-quality telephone circuit may only require about 16 dB. The required signal-to-noise ratio determines the spacing of the line amplifiers in a line communication system and of the relay stations in a microwave radio-relay link, and is a factor in the minimum transmitter power necessary in a radio system. For economic reasons, therefore, it is necessary to maximize the signal-to-noise ratio of a system by reducing the magnitudes of any noise sources as far as possible.

EXAMPLE 9.4

An amplifier has a gain of 30 dB and generates a noise power, referred to its input terminals, of 3 μW. If the signal applied to the amplifier input is −10 dBm† with a signal-to-noise ratio of 20 dB, calculate the signal-to-noise ratio at the output of the amplifier.

Solution
Output signal level = −10 + 30 = +20 dBm
Input noise level P_N = −10 − 20 = −30 dBm

Therefore

$$-30 = 10 \log_{10} \left(\frac{P_N}{1 \times 10^{-3}} \right)$$

$$3 = \log_{10} \left(\frac{1 \times 10^{-3}}{P_N} \right)$$

Taking antilogs, $1000 = \dfrac{1 \times 10^{-3}}{P_N}$, so that

$$P_N = 1 \ \mu W$$

The total amplifier noise, referred to the input, is 3 μW, and hence the total input noise power is 4 μW. In dBm,

$$x = 10 \log_{10} \left(\frac{4 \times 10^{-6}}{1 \times 10^{-3}} \right) = 10 \log_{10} 4 \times 10^{-3} = -24 \text{ dBm}$$

The output noise power is thus −24 + 30 = +6 dBm and the output signal-to-noise ratio is

$$20 - 6 = 14 \text{ dB} \qquad (Ans.)$$

Exercises

9.1. Write a brief account of the main sources of interference to h.f. radio reception, distinguishing carefully between natural and man-made interference. State briefly what steps may be taken to combat each type of interference. (C&G)

9.2. Briefly explain the origin of thermal noise. What is meant by the term *white noise*?

An amplifier has a rectangular bandpass characteristic from 300 kHz to 500 kHz and a gain of 40 dB. Calculate the noise voltage at its output due to a 100 kΩ resistor at 20°C connected across its input. What will be the change in output noise if the resistor temperature is reduced to −190°C?

Boltzmann's constant is 1.38×10^{-23} J/° and absolute zero temperature (0 °) is −273°C.

9.3. Explain briefly the main sources of internal noise in either a transistor or a valve used in a high-frequency amplifier. What is meant by the term signal-to-noise ratio?

Determine the signal-to-noise ratio in a receiving system in which the received signal level is 10 μW and the effective noise power at the receiver input is given by $N = kTB$ watts, where k is a constant = 1.38×10^{-23} W/°/Hz, T = the equivalent noise temperature of the input circuit (°), and B = the effective bandwidth (Hz). The effective bandwidth of the receiver is given as 200 kHz and the effective noise temperature is 3000 °. Give your answer in decibels. (C&G)

† dBm: decibels relative to 1 milliwatt.

9.4. Discuss the possible sources of noise in a coaxial-line transmis sion system.

The noise power in one channel of a coaxial-line system i −138 dBm at the input to the first repeater. If the noise due t the preceding terminal equipment is negligible, what channe power is required from the latter to produce a signal-to-nois ratio of 62 dB if the signal attenuation through the line i 39 dB?

Assuming that the noise in the amplifier is uniformly distri buted over the frequency band, how would the signal-to-nois ratio vary with frequency over the amplifier bandwidth, if th signal at the output of the terminal equipment were maintaine at a constant level? (C&G

9.5. (a) What is meant by *noise* in a communication circuit?

(b) Tabulate the main sources of noise, both external an internal, which are common in communication receivers, an indicate briefly methods of suppressing or reducing them.

(c) If the effective noise level at the input of a receiver i −130 dBm, and the signal-to-noise ratio expected is 23 dB what receiver gain is required to produce an output of 25 mW? (C&G

9.6. (a) Explain briefly the following types of noise occurring i low-level amplifiers: (i) shot, (ii) partition, (iii) flicker, (iv microphonic, (v) thermal agitation. (b) Which noise effects ar most significant in (i) audio-frequency amplifiers, (ii) high frequency amplifiers? (C&G

9.7. (a) Describe the types of noise in semiconductors which ar caused by (i) shot effect, (ii) flicker effect ($1/f$) noise, (iii microphony, (iv) partition. (b) Explain briefly the difference between the above types of noise and that produced by light ning or spark discharge. (c) The noise produced at the input of receiver is given by 100 kTB watts. Calculate the level of signa required at the input to provide a signal-to-noise ratio of 30 dE for the following conditions: T, the absolute temperature = 290°; B, the bandwidth = 10 kHz; $k = 1.38 \times 10^{-23}$ J/° (C&G

Short Exercises

9.8. Why is a f.e.t. less noisy than a bipolar transistor?

9.9. Will shot noise in a bipolar transistor increase or decrease if th d.c. collector current of the transistor is reduced? Give a reaso for your answer.

9.10. The thermal noise power at the input of a radio-receiver i 10 μW. What will be the noise power if (i) the noise tempera ture is doubled, (ii) the bandwidth is doubled, (iii) both th noise temperature and the bandwidth are doubled?

9.11. List the sources of noise external to a radio receiver.

9.12. List the ways in which the effects of external noise can b minimized.

9.13. Define the term signal-to-noise ratio. Why cannot signal-to noise ratio be measured? What parameter is normally meas ured in its place?

9.14. The signal-to-noise ratio at the output of a system is 33 dB. I the signal power is 50 mW determine the noise power.

9.15. Calculate the r.m.s. noise voltage developed in a 330 kΩ resis tor in a 2 MHz bandwidth at room temperature.

10 Integrators and Differentiators

Introduction

Many circuits in electronic, radio and television equipment rely for their correct operation upon the *integration* or the *differentiation* of an input waveform. The input waveform may be sinusoidal or rectangular or perhaps may represent a speech, video or a music signal. The action of an *integrator* is to perform electrically the mathematical operation of integration and similarly a differentiator carries out the mathematical process of differentiation on an input waveform.

Differentiation

When a voltage v is differentiated the rate of change dv/dt of that waveform is obtained. Suppose, for example, that a sinusoidal voltage, $v = V \sin \omega t$, is to be differentiated. Fig. 10.1a shows the voltage waveform. When the voltage is zero and about to go positive, its rate of change has its maximum positive value; hence at this moment the differentiated wave is at its maximum positive value, see point t_1 on Figs. 10.1a and b. As the voltage wave increases towards its maximum positive value, its rate of change decreases and when the positive peak value is reached, point t_2, the rate of change of voltage has fallen to zero. The voltage now falls to zero at time t_3 before rising to its negative peak value at t_4. The rate of change of voltage over the time period $t_2 \rightarrow t_4$ is negative with the maximum rate of change occurring at t_3. After time t_4 the voltage wave falls from its peak negative value and reaches zero volts at t_5 and then rises positively to reach its positive peak value at t_6. Thus the rate of change of voltage is positive over the period t_4 to t_t with its maximum value at t_5. The rate of change of voltage waveform is shown in Fig. 10.1b and is clearly cosinusoidal.

For another example consider the square voltage waveform shown in Fig. 10.2a. At time t_1 the voltage rises suddenly from zero to its maximum positive value of V volts. The rate of change of voltage is very large if the rise time of the pulse is small. In the time interval t_1 to t_2 the voltage remains constant and so its rate of change is zero. At time t_2 the voltage abruptly falls from $+V$ volts to $-V$ volts and then remains constant at this value until time t_3. The rate of change of voltage is negative and large at time t_2 but is zero for the period t_2 to t_3. At time t_3 the voltage suddenly changes from $-V$ to $+V$ volts and the rate of change of voltage is positive and of magnitude $2V/$(time taken for change to occur). From t_3 to t_4 the voltage is constant at $+V$ volts and so its rate of change is zero. This means that differentiation of a square waveform produces a number of equally spaced positive and negative spikes as shown in Fig. 10.2b.

Integration

Integration is the reverse process to differentiation. Hence if the waveforms shown in Figs. 10.1b and 10.2b are integrated the resultant waveforms are illustrated by Figs. 10.1a and 10.2a.

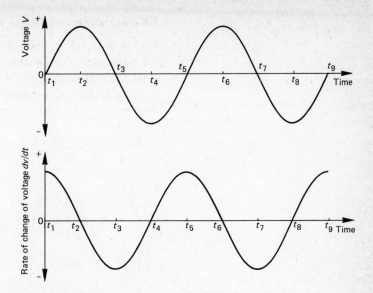

Fig. 10.1 Differentiation of a sinu-soidal voltage wave

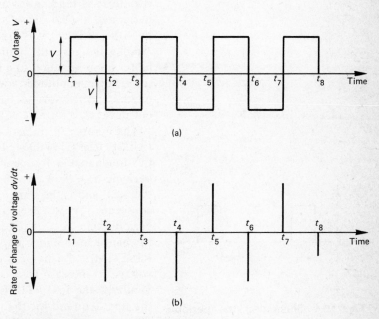

Fig. 10.2 Differentiation of a square voltage wave

Characteristics of Pulse Waveforms

Fig. 10.3 shows a sinusoidal voltage, known as the fundamental frequency, and another, smaller amplitude voltage at three times the fundamental frequency and known as the third harmonic. At time $t = 0$ the two voltages are in phase with one another but, of course, will not remain so. The waveform

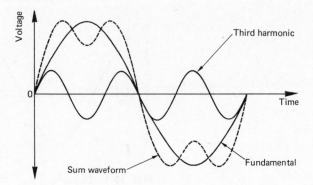

Fig. 10.3 Waveform produced by a fundamental and its third harmonic

produced by summing the instantaneous values of the two voltages is shown dotted; it is clear that the resultant waveform is tending towards a rectangular shape. If the fifth harmonic, also in phase at time $t = 0$, is added to the dotted line, the resultant waveform becomes even more rectangular. Adding the seventh, ninth, etc. harmonics produces an even better approximation to a rectangular waveshape. By changing the relative amplitudes and phases, at time $t = 0$, of the fundamental and the various harmonics, both odd and even, different pulse waveforms can be obtained.

The number of pulses occurring per second is known as the PULSE REPETITION FREQUENCY (p.r.f.) and is equal to the fundamental frequency contained in the waveform. The periodic time T of a rectangular waveform is the reciprocal of the p.r.f. and is the time interval between the leading edges of consecutive pulses (see Fig. 10.4). Also, in Fig. 10.2a the periodic time is $t_3 - t_1$. It is not possible for a pulse waveform to change instantaneously from one value to another and the RISETIME of a pulse is defined as the time t_r taken for the voltage to increase from 10% to 90% of its maximum value. Similarly, the FALLTIME or DECAY TIME t_f of a pulse is the time required for the pulse amplitude to fall from 90% to 10% of its maximum value. The WIDTH or DURATION of a pulse is normally taken as the distance, along the time axis, between the 50% maximum amplitude points on the leading and trailing edges of the pulse. Risetime and falltime and pulse width are illustrated by Fig. 10.5. When a pulse is transmitted through a system having inadequate gain at low frequencies the pulse may also exhibit *sag* (see Fig. 10.6).

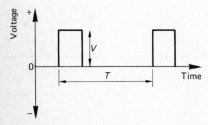

Fig. 10.4 Illustrating the periodic time and amplitude of a pulse waveform

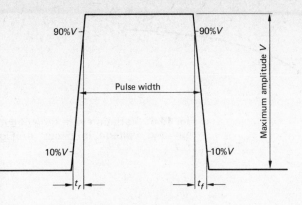

Fig. 10.5 Illustrating the risetime, falltime, and width of a pulse

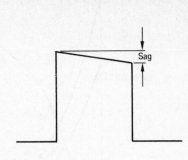

Fig. 10.6 Illustrating the sag of a pulse

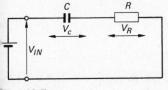

Fig. 10.7

Transient Response of *R-C* Circuits

Fig. 10.7 shows a simple circuit consisting of a resistor R connected in series with a capacitor C. At any instant in time the direct voltage V_{IN} applied to the circuit is equal to the sum of the voltages V_R and V_C appearing across the two components. When the input voltage is first applied, the capacitor is discharged and all of the voltage will appear across the resistor. The initial current I_0 that flows into the circuit is equal to the input voltage divided by the circuit resistance and has the maximum possible current value. The current flowing charges the capacitor and as its voltage rises the voltage across the resistor must correspondingly fall. The variations with time of current and voltage in the circuit are exponential in nature and are given by the following equations:

$$i = (V_{IN}/R)e^{-t/CR} = I_0e^{-t/CR} \tag{10.1}$$

$$v_R = V_{IN}e^{-t/CR} \tag{10.2}$$

$$v_C = V_{IN}(1 - e^{-t/CR}) \tag{10.3}$$

Graphs illustrating the variation of current and voltage with time are given in Fig. 10.8. The product CR has the dimensions of seconds and is known as the TIME CONSTANT of the circuit. The time constant is the time in which the current in the circuit would fall to zero if the initial rate of decrease were maintained. However, immediately the capacitor starts to charge, its voltage V_C increases and so the current $(V_{IN} - V_c)/R$ is reduced.

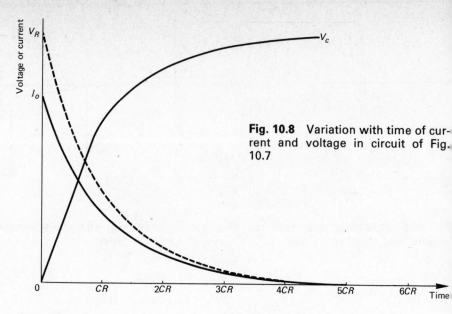

Fig. 10.8 Variation with time of current and voltage in circuit of Fig. 10.7

When the time t is equal to the time constant, then $i = I_0e^{-t/CR}$ becomes

$$I_0 = e^{-CR/CR} = I_0e^{-1} = 0.368I_0$$

$$V_R = 0.368V_{IN} \quad \text{and} \quad V_c = (1-0.368)V_{IN} = 0.632V_{IN}$$

In the next CR seconds i and V_R fall to $0.368 \times 0.368 = 0.135$ times their initial values whilst V_C increases by 0.632 of the difference between V_{IN} and the value it had reached after the first CR seconds. That is, after $2CR$ seconds

$$V_C = 0.632V_{IN} + 0.632 \times (V_{IN} - 0.632V_{IN})$$

$$= 0.632V_{IN} + 0.632V_{IN} \times 0.368 = 0.865V_{IN}$$

In the following CR seconds i and V_R fall to $0.368 \times 0.135 = 0.05$ times their initial value; V_c rises to

$$0.865V_{IN} + 0.632(V_{IN} - 0.865V_{IN}) = 0.95V_{IN}$$

Similarly, after $4CR$ seconds i and V_R are equal to 0.018 times their initial value and $V_C = 0.982V_{IN}$ and so on. After $5CR$ seconds the current has fallen to very nearly zero and the capacitor voltage has almost reached its maximum value of V_{IN} volts. In practice, it is usual to assume the transient is completed after $5CR$ seconds.

EXAMPLE 10.1

A constant direct voltage of 10 V is maintained across a circuit consisting of a 10 kΩ resistor connected in series with a 0.1 μF capacitor. Draw the variations with time of the voltages across the capacitor and the resistor. From the graphs determine the risetimes or falltimes of the voltages.

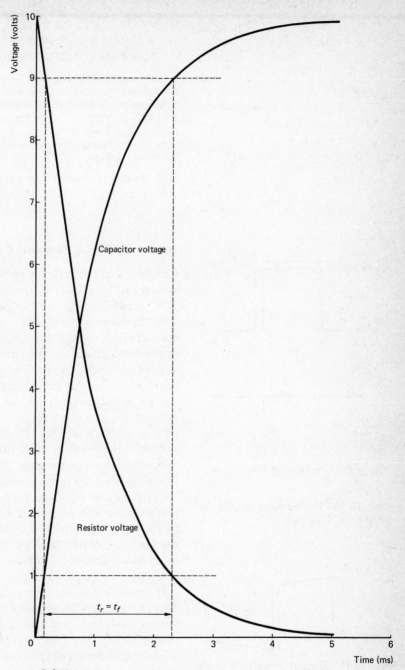

Fig. 10.9

Solution
The time constant of the circuit is $10^4 \times 10^{-7} = 1$ millisecond. The initial values of the capacitor and resistor voltages are, respectively, 0 volts and 10 volts. After 1 ms,

$$V_R = 10 \times 0.368 = 3.68 \text{ volts} \quad \text{and} \quad V_C = 10 \times 0.632 = 6.32 \text{ volts}$$

1 ms later

$$V_R = 0.368 \times 3.68 = 1.354 \text{ volts} \quad \text{and}$$
$$V_C = 6.32 + 0.632(10 - 6.32) = 8.646 \text{ volts}$$

After 3 ms

$$V_R = 0.368 \times 1.354 = 0.498 \text{ volts} \quad \text{and}$$
$$V_C = 8.646 + 0.632(10 - 8.646) = 9.502 \text{ volts}$$

Similarly, after 4 ms and then 5 ms have elapsed $V_R = 0.183$ volts and then 0.067 volts while V_C is first equal to 9.817 volts and then to 9.933 volts. A good approximation to the required curves can be obtained by plotting the points calculated above and this has been done in Fig. 10.9.

From the graph the risetime is equal to the falltime and is 2.2 ms. (*Ans.*)

It should be noted that the risetime is 2.2 times the time constant.

Response of *R-C* Networks to Rectangular Pulses

Consider that the perfectly rectangular pulse of width τ seconds shown in Fig. 10.10a is applied to the input terminals of the *R*-*C* network of Fig. 10.10b. Suppose that τ is very much less than the time constant, CR seconds, of the network and that the capacitor is initially discharged. When the pulse is first applied the maximum current will flow in the circuit and the voltage across R, which is the output voltage, will jump to the same value as the input pulse. The capacitor is unable to charge instantaneously and at the moment the pulse is first applied acts as a short-circuit. As the capacitor charges up, the voltage across R will fall but since $\tau \ll CR$ the reduction in the output voltage will be very little at the time the pulse ends. When the input pulse ends, the capacitor will commence to discharge through resistor R with time constant CR seconds. The polarity of the capacitor voltage is shown in the diagram and so the discharge current will flow in the opposite direction to the charging current. Because of this the output voltage of the circuit suddenly falls by $-V$ volts, taking it negative, and then decays towards zero volts with time constant CR seconds. The output voltage waveform is shown in Fig. 10.11.

If a square waveform varying between 0 and $+V$ volts (Fig. 10.12a) is applied to the circuit of Fig. 10.10b, the output voltage waveform will take a few cycles to settle down to a *steady-state* value. Figs. 10.12b and c show how the voltages across the resistor and the capacitor vary when the pulse width τ is very much less than the time constant CR of the circuit. When the first pulse is applied the voltage across R abruptly rises from zero to $+V$ volts and then starts to decay towards zero volts as the capacitor charges up. When the first pulse ends the output voltage has fallen to $Ve^{-\tau/CR}$ volts and rapidly changes negatively by V volts. The output voltage is taken negative and discharges from this value towards zero but

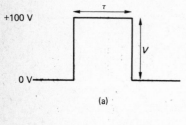

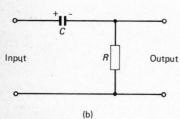

Fig. 10.10 Rectangular pulse applied to *R-C* circuit

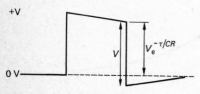

Fig. 10.11 Output waveform of circuit in Fig. 10.10

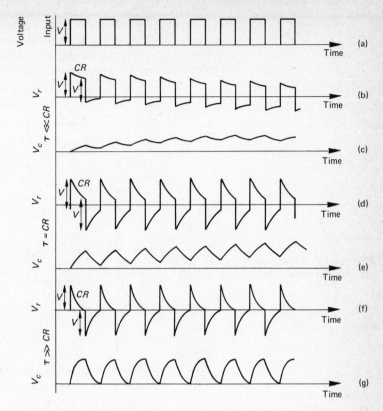

Fig. 10.12 Effect of applying a square waveform to an *R-C* circuit

before it reaches zero the second input pulse is applied. The output voltage now rises in the positive direction by *V* volts but does not reach +*V* volts because it has started from a small negative voltage. As the capacitor charges the output voltage falls exponentially and at the end of the second pulse again falls suddenly by −*V* volts. This procedure is repeated as each input pulse is applied and after a few cycles a nearly square output voltage, centred about the zero voltage line, is obtained. The peak value of the output waveform is +*V*/2 volts. At all times the sum of the output voltage and the voltage across the capacitor is equal to the input voltage; hence the capacitor voltage builds up and eventually stabilizes as a triangular voltage centred about a potential of +*V*/2 volts.

When the time constant is equal to the pulse width the capacitor is able to charge to a much greater extent whilst an input pulse is present. As a result the variation with time of the output voltage is much more pronounced. This is shown by Figs. 10.12*d* and *e* from which it is evident that the output voltage is no longer very nearly of square waveform. If the time constant of the circuit is made much smaller than the pulse width (Figs. 10.12*f* and *g*) the above effect is very pronounced. The capacitor is now able to charge up to the input pulse voltage during the duration of a pulse and this, of course, means that the output voltage falls to zero before pulse ends. Clearly, if an *R-C* circuit is required to pass square waveform it is essential that its time constant shoul~~ not be small compared with the duration of the pulses.

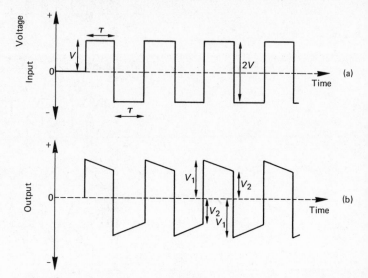

(a)

(b)

Fig. 10.13 Effect of applying a ± square waveform to an *R-C* circuit

A similar result is obtained when a square voltage waveform which varies between ±*V* volts is applied to a series *R-C* circuit. After a few cycles the steady-state output waveform shown in Fig. 10.13 is obtained. The change in the amplitude of the output waveform is equal to the amplitude change at the input. Therefore

$$2V = V_1 + V_2 = V_1 + V_1 e^{-\tau/CR}$$

or

$$V_1 = \frac{2V}{1 + e^{-\tau/CR}} \tag{10.4}$$

The voltage across the capacitor is always equal to the input voltage minus the output voltage.

EXAMPLE 10.2

A ±20 V square waveform is applied to the input terminals of a circuit consisting of a 100 kΩ resistor connected in series with a 0.01 μF capacitor. If the pulse duration is 0.05 ms sketch the output waveform.

Solution

From equation (10.4), $V_1 = \dfrac{40}{1 + e^{-0.05/1}} = 20.5$ V. Therefore

$$V_2 = 40 - 20.5 = 19.5 \text{ V}$$

The required waveform is shown in Fig. 10.14.

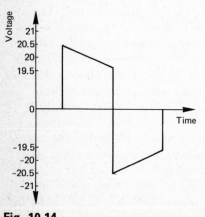

Fig. 10.14

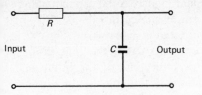

Fig. 10.15 Basic integrating circuit

Integrating and Differentiating Circuits

Fig. 10.15 shows the circuit of a basic integrating circuit. It consists of the series connection of a resistor R and a capacitor C whose values are such that the time constant CR of the circuit is large compared with the periodic time T of the input voltage waveform. The output voltage is taken from the capacitor. The general equation for the circuit is

$$V_{IN} = V_R + V_C \quad \text{but since } CR \gg T,$$

$$V_C \ll V_R \text{ and } V_{IN} \simeq V_R = iR$$

Also,

$$V_{OUT} = V_C = \frac{Q}{C} = \frac{1}{C} \int i \, dt \simeq \frac{1}{C} \int \frac{V_R}{R} \, dt \simeq \frac{1}{C} \int \frac{V_{IN}}{R} \, dt$$

or

$$V_{OUT} = \frac{1}{CR} \int V_{IN} \, dt \tag{10.5}$$

Equation (10.5) states that the output voltage of the circuit is the time integral of the input voltage.

When a square wave varying between 0 and $+V$ volts is applied to the integrator, the capacitor voltage will increase exponentially with time. Since $CR \gg T$, only the straight part of the exponential will have been completed by the time the first pulse ends. The capacitor will then commence to discharge but will not have completely done so by the time the second pulse arrives and so on. In this way the capacitor voltage will build up (see Fig. 10.12c) until after a few cycles a steady state condition is reached.

A better approach to true integration of an input waveform is obtained if an operational amplifier integrator is employed and Fig. 10.16 shows the basic circuit. The inverting terminal is a virtual earth point and so an input current $I_{IN} = V_{IN}/R_1$ flows through R_1 and, since the input impedance of the operational amplifier is high, through C_1. Therefore

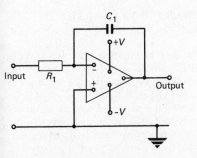

Fig. 10.16 Op-amp integrator

$$V_{OUT} = -\frac{1}{C_1} \int i \, dt = -\frac{1}{C_1} \int \frac{V_{IN}}{R_1} \, dt = -\frac{1}{C_1 R_1} \int V_{IN} \, dt \tag{10.6}$$

Equation (10.6) states that the output voltage is proportional to the time integral of the input voltage. Consider again the resistance capacitance circuit given in Example 10.2. The time constant CR is 20 times larger than the pulse duration of the square waveform (10 times greater than the periodic time) and hence the circuit acts as an integrator if the output is taken from the capacitor. At all times $V_{IN} = V_C + V_R$ hence when

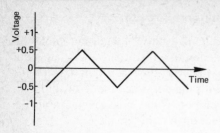

Fig. 10.17 Triangular waveform

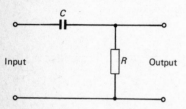

Fig. 10.18 Basic differentiating circuit

$V_R = \pm 20.5$ V then $V_C = \mp 0.5$ V and when $V_R = \pm 19.5$ V then $V_C = \mp 0.5$ V. The voltage appearing across the capacitor therefore has a triangular waveform as shown in Fig. 10.17.

The basic circuit of a differentiator is shown in Fig. 10.18. The component values are chosen so that the time constant CR is short relative to the periodic time of the input signal waveform. The general equations describing the circuit are

$$V_{IN} = V_R + V_C \simeq V_C$$

and

$$i = dq/dt = C\, dV_C/dt \simeq C\, dV_{IN}/dt$$

The output voltage V_{OUT} of the differentiator is developed across the resistor R. Therefore

$$V_{OUT} = iR = CR\, dV_{IN}/dt \qquad (10.7)$$

Equation (10.7) shows that the output voltage is proportional to the differential with respect to time of the input voltage.

The operational amplifier version of a differentiating circuit is easily obtained by interchanging the positions of R_1 and C_1 in the integrator circuit of Fig. 10.16.

Some examples of the waveforms that can be produced by the processes of integration and differentiation are given in Fig. 10.19. Three waveforms, a, d and g, are shown integrated in b, e and h, and differentiated in c, f and i respectively.

Exercises

10.1. Draw a pulse waveform having (i) p.r.f. $= 10$ kHz, (ii) pulse width $= 40\ \mu$S, (iii) sag $= 5\%$, (iv) risetime $= 5\ \mu$s, (v) falltime $= 7\ \mu$s.

10.2. In the circuit of Fig. 10.20, T_1 is normally conducting and the input pulse is of sufficient amplitude to cut off the transistor. Draw to scale the output voltage V_{OUT} during the input pulse period when (a) C is $1.0\ \mu$F, (b) C is $0.1\ \mu$F. (C&G)

10.3. Explain with basic circuits how the waveforms of an electrical signal may be (a) differentiated, (b) integrated, Sketch the output waveform when the input waveform is (i) triangular, (ii) rectangular, (iii) sinusoidal.

10.4. Explain, with circuit diagrams, what is meant by (a) a differentiating circuit and (b) an integrating circuit. Sketch the waveforms of the outputs of these circuits if there is applied to the inputs (i) a square wave, (ii) a triangular wave.

10.5. A train of rectangular pulses is applied to the input of the circuit shown in Fig. 10.21. The pulses have a repetition frequency of 1000 Hz, an amplitude of 100 V, and a duration of $10\ \mu$s. Sketch the waveform of the output voltage.

10.6. Explain what is meant by the terms differentiating circuits and integrating circuits. The waveform shown in Fig. 10.22 is applied to (a) a differentiator, (b) an integrator. Sketch the output waveform in each case. (C&G)

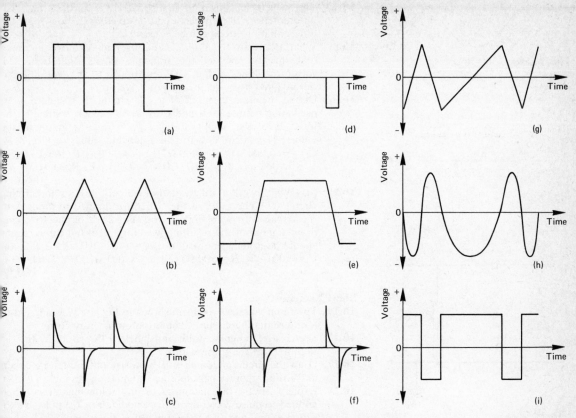

Fig. 10.19 Waveforms produced by integration and by differentiation

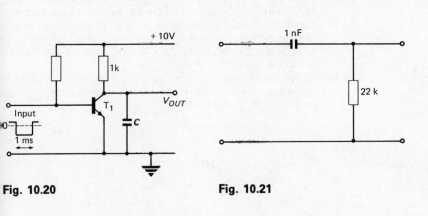

Fig. 10.20

Fig. 10.21

Fig. 10.22

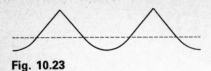

Fig. 10.23

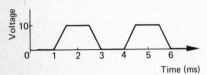

Fig. 10.24

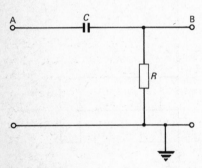

Fig. 10.25

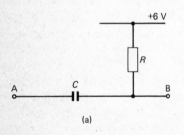

(a)

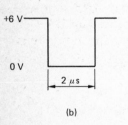

(b)

Fig. 10.26

10.7. The waveform shown in Fig. 10.23 is applied to an ideal differentiator. Sketch the output waveform. Show how the differentiating circuit can be constructed. (C&G)

10.8. What is meant by a differentiating circuit? Give a circuit diagram and explain the principle of its operation. The waveform shown in Fig. 10.24 is applied to an ideal differentiator. Draw to scale the output waveform.

10.9. Define the time constant of a CR circuit. If a square wave oscillating between 0 V and -6 V at 500 kHz is applied to the input A of the circuit of Fig. 10.25, draw a graph of the voltage waveform at B for the following values of C and R; (a) $C = 0.001\ \mu\text{F}$, $R = 500\ \Omega$, (b) $C = 0.001\ \mu\text{F}$, $R = 1\ \text{k}\Omega$, (c) $C = 0.0001\ \mu\text{F}$, $R = 2\ \text{k}\Omega$, (d) $C = 0.001\ \mu\text{F}$, $R = 1\ \text{M}\Omega$. (C&G)

10.10. (a) What is (i) a differentiating circuit, (ii) an integrating circuit? (b) The pulse in Fig. 10.26b is applied to input A of the circuit shown in Fig. 10.26a. For each of the following draw a graph of the resulting waveform at B (all three graphs should be to the same scale): (i) $C = 0.001\ \mu\text{F}$, $R = 250\ \Omega$, (ii) $C = 0.001\ \mu\text{F}$, $R = 500\ \Omega$, (iii) $C = 0.001\ \mu\text{F}$, $R = 1\ \text{k}\Omega$. (C&G)

Short Exercises

10.11. The ramp voltage waveform shown in Fig. 10.27 is applied to a differentiating circuit. Sketch the output waveform.

10.12. Sketch a rectangular pulse and label (i) the risetime, (ii) the falltime, (iii) the pulse width.

10.13. Draw the circuit diagrams of (i) a passive differentiating circuit and (ii) an operational amplifier integrating circuit.

10.14. Explain briefly the meanings of the following terms when applied to pulse waveforms: (i) periodic time, (ii) pulse width, (iii) risetime, and (iv) pulse repetition frequency.

10.15. Deduce and draw the waveform of the capacitor voltage in the circuit shown in Fig. 10.10.

10.16. Draw the circuit diagram of an op-amp differentiating circuit. Prove that your circuit carries out the differentiation process.

10.17. Sketch the output waveform of an integrating circuit when a rectangular pulse is applied to the input when (a) the pulse width is much greater than the CR time, (b) the pulse width is much less than the CR time.

10.18. Sketch the output waveform of a differentiating circuit for a rectangular input pulse whose duration is (a) much longer than the CR time, (b) much shorter than CR time.

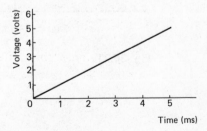

Fig. 10.27

11 Stabilized Power Supplies

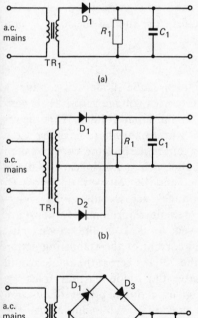

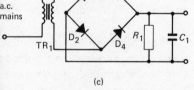

Fig. 11.1 Rectifier circuits (*a*) half-wave, (*b*) full-wave, (*c*) full-wave bridge

Introduction

All electronic equipments require a power supply of some kind to provide the necessary d.c. operating voltages and currents. Some portable equipments, such as transistor radio receivers, are battery operated but the majority of equipment employs an electronic power supply. The basic power supply consists of (1) a transformer whose function is to convert the a.c. mains supply voltage to the lower value required by the equipment, (2) a rectifier unit whose function is to convert the a.c. voltage supplied by the transformer to a d.c. voltage, and (3) a filter whose purpose is to remove *ripple* from the rectified voltage. The rectifier unit may be operated on either a half-wave or a full-wave basis and Fig. 11.1 gives the circuits of the three kinds of rectifier which are most often used. The principles of operation of these circuits have been discussed in [E II].

The correct operation of many equipments demands that the direct power supply voltage is maintained at a constant value, within fairly fine limits, even though the input mains voltage and/or the current taken from the power supply may vary. Generally, the inherent regulation of a supply is inadequate to meet the demands placed upon it by the supplied equipment and then some kind of voltage stabilization circuitry must be provided. The function of a voltage stabilizer is to maintain a constant voltage across the load as the input voltage and/or the load current vary within specified limits.

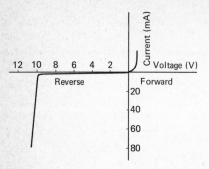

Fig. 11.2 Zener diode characteristic

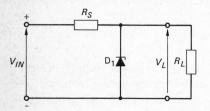

Fig. 11.3 Zener diode stabilizer

The Zener Diode Stabilizer

The zener, or voltage reference, diode is a semiconductor device which has a normal diode current voltage characteristic in the forward direction; in the reverse direction the current abruptly increases if the voltage across it reaches a critical value known as the breakdown voltage. The voltage across the diode will thereafter remain more or less constant even though the current through the diode may vary considerably. A typical zener diode current voltage characteristic is shown in Fig. 11.2.

The zener diode can be used to stabilize the output voltage of a power supply since large changes in the current flowing in the diode produce only small changes in the voltage across the diode.

The circuit of a simple zener diode voltage stabilizer is shown in Fig. 11.3. The voltage V_L across the load is equal to the input voltage V_{IN} minus the voltage dropped across the series resistor R_S; this is equal to the sum of the load I_L and diode I_d currents times R_S, i.e.

$$V_{IN} = (I_L + I_d)R_S + I_L R_L \qquad (11.1)$$

If the input voltage should increase, the diode will conduct a greater current so that the increase in voltage across R_S is very nearly equal to the increase in the input voltage. This means that the load voltage does not vary to anywhere near the same extent as does the input voltage. Similarly if the current taken by the load should increase, the current flowing in the diode will fall by the same percentage and the voltage drop across R_S will not change. The voltage across the load $V_L = V_{IN} - (I_d + I_L)R_S$ will hence remain approximately constant. The minimum current allowed to flow in the diode must always exceed the breakdown current or the stabilizing action of the diode will be lost, i.e. the voltage across the diode would vary with current. Conversely, the diode current must be limited to the value which gives a diode power dissipation which does not exceed the maximum value quoted by the manufacturer.

If the stabilizing circuit will be called upon to stabilize changes in output voltage caused by a varying load current only, the value of R_S can readily be determined by transposition of equation (11.1) to give

$$R_S = \frac{V_{IN} - V_L}{I_L + I_d} \qquad (11.2)$$

If, however, changes in the input voltage may also be expected, the value of R_S must be calculated to ensure satisfactory operation with the most adverse combination of input voltage and load current [E II].

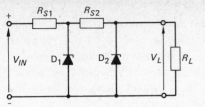

Fig. 11.4 Double zener diode stabilizer

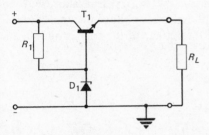

Fig. 11.5 Emitter-follower stabilizer

The voltage stability of the basic zener diode circuit is not good enough for many applications because the internal resistance of the diode is not zero. This means that any change in the current flowing in the diode will cause a small change in the voltage appearing across the diode and hence also in the output voltage. The stability of the output voltage can be improved by cascading two zener diodes as shown by Fig. 11.4.

The stabilization efficiency of the voltage regulator would be increased if the magnitude of the current flowing in the diode were reduced. The circuit of an emitter-follower voltage stabilizer is shown in Fig. 11.5. Since the transistor is connected as an emitter follower, the voltage at its emitter, which is the load voltage, is very nearly equal to the base voltage. The base voltage is specified by the zener diode and so the output voltage is held constant within limits determined by the diode characteristics. The emitter-follower stabilizer has no provision for varying the output voltage and its stabilization efficiency is not good enough for many applications.

Series-Control Stabilizers

The principle of a much more efficient type of stabilizer is shown in block schematic form by Fig. 11.6.

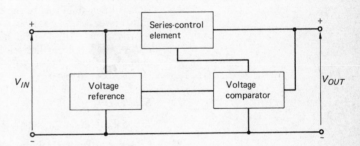

Fig. 11.6 Block schematic diagram of series stabilizer

The output voltage of the rectifier is applied to a series-control element which introduces resistance into the positive supply line. The output voltage V_{OUT} is smaller than the input voltage by the voltage dropped across the series element. The output voltage, or a known fraction of it, is compared in the voltage comparator with a voltage reference. The difference between the two voltages is detected and an amplified version of it is applied to the series-control element in order to vary its resistance in such a way as to maintain the output voltage at its correct value. If, for example, the output voltage is larger than it should be, the amplified difference voltage will be of such a polarity that the resistance of the controlled element will be made larger and the output voltage will fall. Conversely, if the

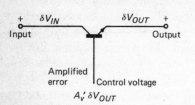

Fig. 11.7 Transistor series-control element

output voltage is less than its correct value, the resistance of the series-controlled element will be reduced by the amount necessary for the output voltage to rise to its correct value. Generally, the series-control element is a transistor connected as shown in Fig. 11.7. When an n-p-n transistor is employed, its collector is connected to the input terminal, and its emitter is connected to the output terminal, of the circuit since the former is more positive. If the output voltage of the stabilizer should vary by an amount δV_{OUT}, the control voltage appearing at the base terminal of the series transistor will be $A_v' \delta V_{OUT}$, which is the amplified error voltage produced by the comparator. The base–emitter voltage of the transistor is then

$$A_v' \delta V_{OUT} - \delta V_{OUT}$$

and will produce a voltage

$$A_v''(A_v' \delta V_{OUT} - \delta V_{OUT})$$

between the base and collector terminals where A_v'' is the voltage gain of the series transistor. The voltage across this transistor is also equal to $\delta V_{IN} - \delta V_{OUT}$ and therefore

$$\delta V_{IN} - \delta V_{OUT} = A_v''(A_v' \delta V_{OUT} - \delta V_{OUT}) \qquad (11.3)$$

EXAMPLE 11.1

In a voltage stabilizer of the type shown in Fig. 11.6 the error voltage gain of the comparator is -100 and the voltage gain of the series transistor is -10. The rectifier circuit connected to the input terminals of the stabilizer has an output resistance of $200\,\Omega$. Calculate the change in the output voltage that occurs when the load current changes by 20 mA.

Solution

$$\delta V_{IN} = \delta I_L R_{OUT} = 20 \times 10^{-3} \times 200 = 4 \text{ V}$$

Hence, substituting into equation (11.3),

$$4 - \delta V_{OUT} = -10(-100\delta V_{OUT} - \delta V_{OUT})$$
$$4 = 1000\delta V_{OUT} + 10\delta V_{OUT} + \delta V_{OUT}$$
$$\delta V_{OUT} = \frac{4}{1011} = 3.966 \text{ mV} \qquad (Ans.)$$

(1) Fig. 11.8 shows the circuit of a voltage stabilizer in which T_1 is the series control element, T_2 is the voltage comparator, and the voltage reference is provided by the zener diode D_1. The emitter potential of T_2 is maintained at a very nearly constant value by the zener diode D_1, whilst its base is held at a fraction of the output voltage by the potential divider R_2, R_3 and R_4. The difference between the base and emitter potentials is amplified and the amplified error voltage is applied to the base of T_1 to vary the bias voltage provided by resistor R_1.

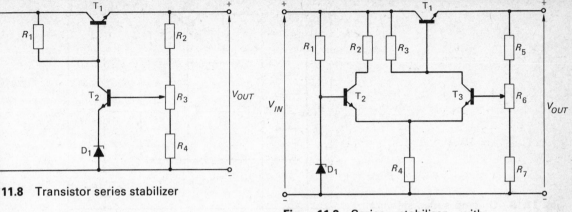

11.8 Transistor series stabilizer

Fig. 11.9 Series stabilizer with differential comparator

Suppose that the output voltage of the stabilizer should increase above its nominal value (set by variable resistor R_3). The base voltage of T_2 will become more positive, with respect to the constant emitter voltage, and T_2 will conduct a larger collector current. The voltage dropped across R_1 will then increase and this will make the base potential of T_1 less positive. T_1 will now conduct less readily and so its resistance increases. The consequent increase in the collector-emitter voltage of T_1 causes the output voltage to fall by an amount that is very nearly equal to the original increase. The series transistor T_1 must be capable of carrying the full load current of the stabilizer and should have an adequate power rating.

(2) An alternative arrangement that is commonly used is given in Fig. 11.9. The voltage comparator is the *differential* amplifier formed by transistors T_2 and T_3. The differential amplifier, or *long-tailed pair*, produces a change of voltage at the collector of T_3 which is proportional to the difference between the base potentials of T_2 and T_3. The base of T_2 is held at a constant voltage by the potential divider formed by R_1 and D_1, while the base voltage of T_3 is a fraction, determined by R_5, R_6 and R_7, of the output voltage. The circuit acts in a similar manner to the previous stabilizer to maintain the output voltage at a more, or less, constant value when the load current is varied. The voltage stability is superior since a greater amplified error voltage is applied to the series transistor when the output voltage is varied, but, on the other hand, the circuit is more expensive in its use of components. The symmetrical nature of the differential amplifier means that the effect on the circuit operation of changes in temperature is reduced, especially if T_2 and T_3 are matched and mounted on a common heat sink.

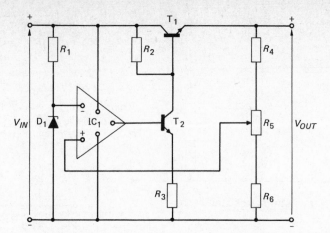

Fig. 11.10 Op-amp series stabilizer

(3) If the amplification provided by the feedback loop is mad
larger the control voltage for a given change in output voltag
will be increased and this will improve the voltage stability o
the circuit. The increase in gain can be provided in a numbe
of different ways and one method which uses an operation:
amplifier is shown in Fig. 11.10. A fraction of the outpu
voltage is applied to the non-inverting terminal of the opera
tional amplifier and the inverting terminal is held at a constar
potential by R_1 and D_1. The amplified difference voltage i
applied to the base of T_2 to vary its collector current an
thereby vary the voltage dropped across R_2. If the outpu
voltage should rise above its nominal value, T_2 will be cause
to conduct harder and so the base potential of T_1 will becom
less positive. T_1 will conduct less readily and an increase
voltage will be dropped between its collector and emitte
terminals which restores the output voltage towards its correc
value.

Shunt-Control Stabilizers

The disadvantage of the series-control stabilizer is that th
control transistor carries the full load current and must be
transistor with a power rating of perhaps several watts. A
alternative which avoids this difficulty is to connect the contro
transistor in parallel with the load. The block schematic diag
ram of a shunt-control stabilizer is shown in Fig. 11.11. If th
output voltage should rise above its nominal value, the resis
tance of the shunt-control is reduced allowing it to pass
larger current. The total current flowing through the serie
resistor R_S is increased and the voltage dropped across it rise
to reduce the output voltage. The shunt-control element mus
be able to handle the maximum output voltage but it does nc
need to be able to carry the full load current. The circuit c
one possible shunt stabilizer is given in Fig. 11.12.

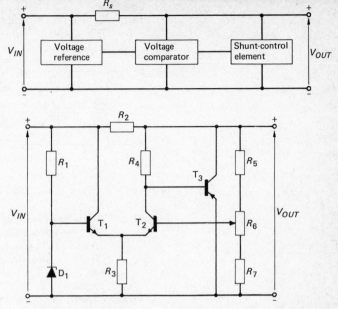

Fig. 11.11 Block schematic diagram of a shunt stabilizer

Fig. 11.12 Transistor shunt stabilizer

The shunt stabilizer is often used for low voltage applications in which a fairly high and constant load current is to be provided. Generally, the series stabilizer is preferred for high voltage, for low or medium load current, and for variable load applications.

Integrated Circuit Voltage Stabilizers

A range of integrated circuit voltage stabilizers, generally known as voltage regulators, are listed in manufacturer's catalogues. These are three-terminal devices which will reduce a large input voltage variation to a much smaller change in output voltage. Typical figures are quoted below for three particular types of regulator:

(a) LM 309 : V_{IN} 7–35 V, V_{OUT} 4.8–5.2 V
(b) LM 340–12 : V_{IN} 14–35 V, V_{OUT} 11.5–12.5 V
(c) LM 340–24 : V_{IN} 26–40 V, V_{OUT} 23–25 V

Fig. 11.13 shows how the LM 309 would be connected to produce a stabilized output voltage of +5 V nominal.

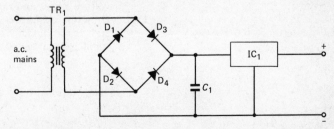

Fig. 11.13 Integrated circuit voltage stabilizer

Exercises

11.1. Explain the operation of the power supply circuit shown in Fig. 11.14.

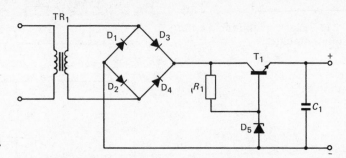

Fig. 11.14

11.2. The voltage stability of a series stabilizer can be improved if the reference voltage is derived from the output voltage instead of from the input voltage. (*a*) Give reasons for this. (*b*) Draw the circuit of an operational amplifier stabilizer which uses this technique. (*c*) Explain the operation of the circuit.

11.3. Explain the operation of the circuit given in Fig. 11.10 when the input voltage varies.

11.4. Explain the operation of the series stabilizer given in Fig. 11.9 when the input voltage changes from its nominal value.

11.5. Explain the operation of the transistor shunt stabilizer shown in Fig. 11.12.

11.6. Fig. 11.15 shows the circuit diagram of a power supply (*a*) List the function of each component. (*b*) Explain the operation of the circuit.

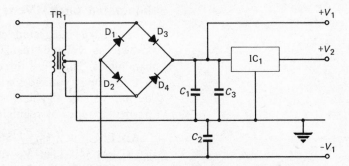

Fig. 11.15

11.7. Fig. 11.16 shows the circuit of a stabilized power unit. Explain the operation of the circuit (*a*) when the input voltage is constant and the load current varies, (*b*) when the input voltage varies and the load current is constant.

11.8. Draw the circuit diagram of a full-wave rectifier (not bridge) and zener diode stabilizer suitable for the provision of a stabilized voltage of negative polarity with respect to earth.

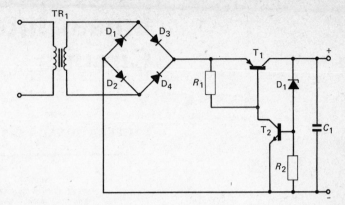

Fig. 11.16

Short Exercises

11.9. What are the requirements for maintaining a constant voltage output across a load?

11.10. Sketch the block diagram of a series stabilized power supply.

11.11. Sketch and explain the block diagram of shunt stabilized power supply.

11.12. What is meant by the term voltage comparator when applied to a stabilized power supply? Draw the circuit of a comparator suitable for use with a stabilized output voltage of negative polarity.

Appendix A
Linear Integrated Circuits

1 OPERATIONAL AMPLIFIERS

	Voltage gain (dB)	Input resist-ance (Ω)	CMRR (dB)	Slew rate (V/μs)	Unity gain band-width (MHz)	Full power band-width (kHz)	Input offset voltage (mV)	Input offset current (nA)
LM 301	104	2 M	90	0.4	1	10	2	3
709	93	250 k	90	0.25	5	200	2	100
741	104	1 M	90	0.5	1	10	1	30
747	104	1 M	90	0.5	1	10	1	30
CA 3140 (mosfet input)	100	1.5×10^9	90	9	4.5	100	5	5×10^{-4}
LF 351 (j fet input)	88	1×10^9	100	13	4	100	5	25×10^{-3}

2 AUDIO-FREQUENCY POWER AMPLIFIERS

	Voltage gain (dB)	Input resist-ance (Ω)	Maximum output power (W)	Band-width (kHz)	% distor-tion	Sensi-tivity (mV)	Power dissipa-tion (W)	Heat sink needed
LM 377	34	3 M	2.5	50	0.1	100	9	Yes
LM 380	34	150 k	5	100	0.2	100	10	Yes
TBA 810	37	5 M	6	20	0.3	75	5	Yes
TBA 820	34	5 M	2	20	0.4	60	1	No

Note The % distortion figures are not obtained at the same time as the maximum output power.

3 RADIO-FREQUENCY AMPLIFIERS

	Voltage gain (dB)	Noise figure (dB)	Input impedance (Ω)	3 dB band-width (Hz)	AGC range (dB)
CA 3002	19	8	100 k	11 M	60
CA 3011/2	60	—	50 k	—	—
CA 3005	16	7.8	—	100	60

4 WIDEBAND AMPLIFIERS

	Voltage gain (dB)	3 dB band-width (MHz)	Input impedance (kΩ)	Noise figure (dB)
CA 3020	75	8	1	—
CA 3023	50	10	—	8.5
CA 3040	34	40	150	9

5 DIFFERENTIAL AMPLIFIERS

	Input offset voltage (mV)	Differential voltage gain (dB)	CMRR (dB)	Input impedance (kΩ)	Input offset current (μA)	3 dB band-width (MHz)
CA 3000	5	28	70	70	—	—
CA 3028	5	39	—	—	6	7
CA 3054	5	32	100	—	2	—

Numerical Answers to Exercises

2.4. 15 mil, 30 mil

2.16. 1 kΩ, 250 Ω

3.1. 0.02 mA, 49.

3.5. 6, 120, 0.5 V, 0.525 V

3.14. 3, $\simeq 1$ MΩ

3.18. 9 mV, 10

3.19. 100

3.23. 18.12

3.29. 255 Hz

4.3. 29.69 dB

4.5. 0.052 dB

4.7. 18.29 dB

4.9. 45×10^4, 2.25×10^4, 0.063

4.12. $R_3 = 4230$ Ω, $R_4 = 42.3$ kΩ

5.6. 5.2 : 1, 465 mW, 1.3 W, 1.74 W, 26.72%

5.8. 1.8 W, 5.25 W, 34.28%, 3.45 W

5.9. 8.08 W

5.17. 11.36 W

5.18. 849 mW

5.20. 14.06 W

6.4. 8000

6.6. 159.2 kΩ

6.10. 0.11 V, 31.3 nH, 313 nH

9.2. 1.793 mV, 0.8391 mV

9.3. 90.82 dB

9.4. 0.2 μW

9.5. 121 dB

9.7. 4 pW

9.10. 20 μW, 20 μW, 40μW

9.14. 25.1 μW

9.15. 102.8 μV

Electronics III: Learning Objectives (TEC)

(3)   *Analyses the circuit applications of fets.*

46   3.1   Predicts the stage gain of a fet common source amplifie stage using a resistive load.

3.2   Outlines the performance of a common source amplifi with

54   *a*) an inductive load,

111   *b*) a tuned circuit load.

32   **(B)**   **Amplifiers**

(4)   *Analyses the performance of voltage amplifiers.*

37, 97, 119   4.1   Explains the biasing conditions for Class A, B, AB ar C operation in the common emitter and common sour modes.

32, 90, 97, 109, 119   4.2   Lists and explains applications of each type of amplifi in 4.1.

48, 51   4.3   Estimates for two-stage class A common emitter ar common source amplifiers: voltage gain, current gai input resistance, output resistance and bandwidth.

4.4   Describes the following types of interstage coupling:

48, 51   *a*) resistance-capacitance,

52   *b*) direct,

112, 114   *c*) transformer.

4.5   Lists applications of the coupling methods stated in 4.4.

58, 59   4.6   Measures the frequency response of the circuits d scribed in 4.3.

57   4.7   Measures signal amplitude limits of the amplifiers d scribed in 4.3.

69   4.8   Measures the effect on the stage gain and bandwidth disconnecting the emitter/source bypass capacitor.

58   4.9   Measures the input and output impedances of the tv stage amplifiers described in 4.3.

111   4.10   States the functions of individual components present an r.f. amplifier.

109, 112   4.11   Explains the selectivity of a tuned amplifier.

135   4.12   Outlines the applications of buffer amplifiers.

4.13   States that for maximum efficiency impedance-matchi must be used in amplifier systems.

(5)   *Understands the function of power amplifiers.*

5.1   Explains the reasons for using power amplifiers in

90   (i) a.f. output stages

119   (ii) r.f. output stages

99   (iii) industrial control systems.

22, 103   5.2   Outlines the relative advantages of both discrete comp nent and integrated circuit power amplifiers.

90, 96, 99, 120   5.3   Identifies from given circuit diagrams four different co figurations of power amplifier.

90, 96, 99, 120   5.4   Explains the primary functions of the main componen in the circuits referred to in 5.3.

(C) Noise

164	(6)	*Applies the basic concepts of electrical noise and its relation to signal strength.*
164	6.1	Defines noise as any unwanted signal.
164, 166, 168, 169, 171	6.2	Lists and explains the sources of noise.
168, 170	6.3	Identifies the precautions taken to minimise the effects of external noise.
172	6.4	Defines signal-to-noise ratio in an amplifier or receiver.
173	6.5	Calculates signal-to-noise ratio in dB, given signal and noise power.

(D) Feedback

64	(7)	*Analyses the general principles of feedback.*
65, 69, 70, 71	7.1	Draws a block diagram of an amplifier with feedback.
64, 78, 127	7.2	Defines positive and negative feedback.
65, 70	7.3	Derives the general expression for the stage gain of an amplifier with feedback.
	7.4	Explains the effects of applying negative feedback to an amplifier in relation to:
65, 70		*a*) gain,
72		*b*) gain stability,
74		*c*) bandwidth,
75, 76		*d*) distortion and noise,
77		*e*) input and output impedance.

(E) Waveform Generators and Switches

127, 143	(8)	*Understands the characteristics of basic sinusoidal oscillator circuits.*
127	8.1	States that oscillations can be produced by an amplifier with positive feedback.
128, 132	8.2	Explains the operation of (*a*) *L-C* oscillators and (*b*) *R-C* oscillators.
135	8.3	States the factors that effect both the short-term and long-term frequency stability of oscillators.
136	8.4	Explains methods of improving the frequency stability of oscillators, e.g. piezo-electric crystal control.
	(9)	*Understands the characteristics of non-sinusoidal oscillators and switches.*
	9.1	Outlines the different types of non-sinusoidal oscillators and switches, such as:
144		*a*) astable multivibrators
148		*b*) monostable multivibrators
152		*c*) bistable multivibrators
158		*d*) Schmitt trigger
161		*e*) Miller integrator
160		*f*) blocking oscillator.
147, 148, 152	9.2	Explains the need for synchronising and triggering of some types of non-sinusoidal oscillator.

Index

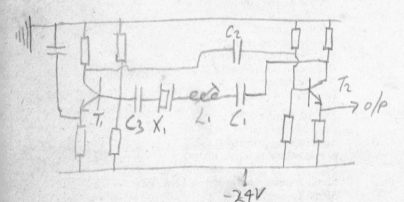

THE CRYSTAL ITSELF IS SERIES RESONANT AT FREQ Fs, BELOW THIS FREQ. IT ACTS CAPACITIVLY WHILE ABOVE THIS FREQ IT ACTS INDUCTIVLY. AT FREQ Fp THE CRYSTAL ACTS INDUCTIVLY BUT FORMS A RESONANT FREQUENCY WITH THE HOLDER CAPACITANCE.

THE CCT WILL OSC. WHEN THE CLOSED LOOP PHASE SHIFT IS 0° & THIS OCCURES ONLY AT THE SERIES RESONANT FREQ OF THE X. ANY LONG TERM FREQ DRIFT CAN BE COMPANSATED BY ADJUSTING L. CLOSED LOOP IS T₁ BASE/COLL. C2. T₂ BASE/COLL. C₁. L₁. X₁. C₃.